Frederick Treves

Scrofula and its Gland Diseases

an introduction to the general pathology of scrofula, with an account of the

histology, diagnosis and treatment of its glandular affections

Frederick Treves

Scrofula and its Gland Diseases
an introduction to the general pathology of scrofula, with an account of the histology, diagnosis and treatment of its glandular affections

ISBN/EAN: 9783337403775

Printed in Europe, USA, Canada, Australia, Japan

Cover: Foto ©berggeist007 / pixelio.de

More available books at **www.hansebooks.com**

AND

ITS GLAND DISEASES

*AN INTRODUCTION TO THE GENERAL PATHOLOGY OF
SCROFULA, WITH AN ACCOUNT OF THE HISTOLOGY, DIAGNOSIS
AND TREATMENT OF ITS GLANDULAR AFFECTIONS*

BY

FREDERICK TREVES, F.R.C.S. Eng.

ASSISTANT-SURGEON TO, AND SENIOR DEMONSTRATOR OF ANATOMY AT, THE
LONDON HOSPITAL; LATE WILSON PROFESSOR OF PATHOLOGY
AT THE ROYAL COLLEGE OF SURGEONS

NEW YORK:
BERMINGHAM & CO., UNION SQUARE.
1882.

W. L. MERSHON & CO.,
Printers, Electrotypers and Binders,
RAHWAY, N. J.

PREFACE.

I BELIEVE I am correct in stating that no special work on the subject of the general pathology of scrofula has appeared in the English language since the publication of the works of Benjamin Phillips, and Robert Glover in the year 1846, and no work on the subject of scrofulous gland disease since the appearance of Dr. Price's monograph in 1861. It must, at the same time, be allowed that scrofula is a disease of some importance, if on no other grounds than those of its frequent occurence and extensive distribution, and that the glandular disorders of the malady form one of the most common and most troublesome affections that come under the notice of the surgeon.

These two facts I offer as some reason for the appearance of this book at the present time.

The works of Phillips and Glover—admirable though they are—were not final as regards our knowledge of strumous processes. Since their publication immense progress has been made in pathological science, a revolution has been effected in the matter of microscopic research, and vast additions have been made to our clinical knowledge and our acquaintance with the life history of disease. In this progress and improvement scrofula has had but a somewhat niggardly share, and the malady, so far as the literature of this country can show us, would appear to have been almost ignored, or dealt with only in a fragmentary manner. Among Continental surgeons and pathologists, however, the subject of scrofulosis has, of late years, received more manifest attention, with the result that the pathology of the disease has been almost reconstructed, its clinical outline more acutely defined, and a more distinct individuality given to the whole disease.

In the present volume I have made extensive use of the valuable material thus provided by the schools of Germany and France. I have endeavoured to give account of the most recent facts that have been brought forward in connection with this wide-spread affection, and the most recent theories that have been expressed as to its nature and relationships.

At the same time I must, in justice to myself, remark that the greater part of the material in this volume is the result of my own investigations into this disease. The clinical facts I have detailed are drawn from a careful examination of a very large number of scrofulous persons, and in such examination I have endeavoured to proceed free from the bias of any preconceived ideas. The opinions also that I have expressed as to the pathological bases of struma are, for the most part, founded upon my own observations; and in the obtaining of material for such observations I have been very fortunate. The account given of the minute changes in the glandular affections is based upon an examination of a great number of glands in various conditions, obtained from more than twenty patients, who, I think, collectively exhibited every aspect of the strumous process. Upon the matter of the treatment of scrofulous gland tumors I have bestowed much attention, and hope that my contributions to this branch of practical surgery will prove of some substantial value. The treatment of strumous gland affections is apt to be a little empirical, a little too exclusive, and, as a rule, very regardless of the fact that one favorite plan of treatment is not necssarily suitable for every phase and condition of a many-sided disease.

I must express my obligations to Dr. T. Smith Rowe, Mr. W. H. Thornton, and Mr. W. Knight Treves, the surgeons to the National Hospital for Scrofula at Margate (where I was for some time resident assistant), for permission to make use of the records of that Institution.

FREDERICK TREVES.

18 GORDON SQUARE:
November 1881.

CONTENTS.

PART I.

THE GENERAL PATHOLOGY OF SCROFULA.

CHAP.		PAGE
I.	INTRODUCTORY.	11
II.	SCROFULA AND TUBERCLE	14
III.	THE NATURE OF TUBERCLE	29
IV.	THE INOCULABILITY OF TUBERCLE	34
V.	A DEFINITION OF SCROFULA	40
VI.	SCROFULA AND PHTHISIS, AND THE ANTAGONISM BETWEEN SCROFULOUS DISEASES	50
VII.	SCROFULA AND ACUTE MILIARY TUBERCULOSIS.	62
VIII.	THE ETIOLOGY OF SCROFULA	64
IX.	THE SCROFULOUS INDIVIDUAL	79

PART II.

SCROFULOUS AFFECTIONS OF THE EXTERNAL LYMPHATIC GLANDS.

	"A SCROFULOUS GLAND"	110
X.	AN OUTLINE OF THE ANATOMY OF THE EXTERNAL LYMPHATIC GLANDS	111

CHAP.		PAGE
XI.	THE ETIOLOGY OF SCROFULOUS LYMPHATIC GLANDS	116
XII.	THE PATHOLOGY OF SCROFULOUS LYMPHATIC GLANDS	126
XIII.	SYMPTOMS AND DIAGNOSIS OF SCROFULOUS LYMPHATIC GLANDS	145
XIV.	THE TREATMENT OF SCROFULOUS LYMPHATIC GLANDS	159

SCROFULOUS LYMPHATIC GLANDS.

PART I.

THE GENERAL PATHOLOGY OF SCROFULA.

CHAPTER I.

INTRODUCTORY.

It must be confessed that the pathology of scrofula is still very ill-defined. Our present knowledge of the disease is encumbered with the unwholesome remains of some centuries of vexed discussion. Since the earliest days of medicine vague ideas and conflicting views appear to have been bestowed upon this disease from time to time, until they have at last formed for it a kind of hereditary property that has slowly accumulated, and has been handed down to the present age with the preciseness of entail. Thus it happens that the scientific limits and outlines of the disease, called scrofula, are blurred and indistinct. Its proper position in the pathological scale is disputed and uncertain. Its relation to other morbid states is variously expressed; and so broken up are opinions as to its nature that, unless a precise definition be given, the term "scrofula" becomes almost an untranslatable word, having a meaning only for the individual who uses it. To what varied conditions of health and to what different individuals does the term "a strumous child" apply! How little some surgeons mean by the term, and how much is implied by its use by others! And when we come to that nebulous person described by the term "slightly strumous," or alluded to as possessing "a touch of scrofula," a sense of utter vacuity is engendered which is almost beyond the reach of scientific relief. No one would think of alluding to a "slightly cancerous person." Cancer exists in a patient or it exists not. It happens to be a distinct disease, and the term "slightly cancerous" would for that reason be ridiculous. Yet the "slightly strumous" are often before us, and they afford no slight indication of what clearness attaches to our present knowledge of the disease.

This lack of scientific limitation in the pathology of scrofula, and this cumbrous heritage of opposed opinions and diverse theories, appear to me to be to a great extent due to two causes, 1. The difficulty of isolating scrofulous disease from the manifestations of mere ill-health, mere frailty of constitution. 2. The persistent attempt of most pathologists to find out some characteristic anatomical element for every disease or diathesis they deal with, and not to remain satisfied until they have found such specific element. The first difficulty is purely clinical, the second has reference to pathological histology.

1. With regard to the clinical difficulty it must be remembered that there is still a wide area in medicine occupied by a class of unhealthy persons, whose morbid state is no more definitely expressed than by saying that they are delicate, of feeble constitution, of frail health. As knowledge advances this area becomes more and more limited. Indeed, it exists but as an evidence of imperfect knowledge and of indistinct notions of morbid change. This condition, known merely as the condition of delicate health, is a blank in medicine; a state of disease without a pathology and without any scientific position. As our acquaintance with disease increases, first one portion and then another of this common and untenanted ground is absorbed, now by one affection, now by another. Some children once classed with the simply delicate are now perhaps known to be the subjects of hereditary syphilis, of ill-defined rickets, or of some hereditary conditions that have now become better known. And so long as there exists a class of individuals whose deviation from the normal state can be expressed in no clearer terms than that they are "deli-

cate" and "of feeble constitution," so long must medical knowledge be considered incomplete. Scrofula has been bounded extensively by this chaotic district, and from time immemorial there has been a reciprocity between them. It is no wonder then if the limits of the disease have been ever confused and fluctuating. In less modern times any chronic state of ill-health was accredited to scrofula, and even now this tendency is not quite extinguished. As Henle* well remarks, "Scrofula is the receptacle into which one vaguely casts all the ailments which afflict children under fourteen years, and of which we do not know the cause." Before hereditary syphilis was understood all its manifestations were classed as scrofulous; rickets also was a strumous disease, as was also chronic hydrocephalus. Lugol,† still more generous, mentions favus, lice, and worms as scrofulus disorders. Carmichael ‡ discovered that scrofula and diabetes were allied; while Hamilton § observes, "I never knew a scirrhus or a cancer take place but in a scrofulous habit."

2. The second cause for the unstable position of scrofula in general pathology depends upon an ancient impression that every disease or diathesis must have some specific anatomical feature associated with it. The outcome of this impression has brought into the field the subject of tuberculosis. Since tubercle was first described its fortunes and those of scrofula have been linked together. In all its changes, in all its losses, in all the false positions into which it has been thrust first by one pathologist and then by another, scrofula has had a share. Scrofula at one time posed as a tubercular process, tubercle at another has been described as a scrofulous process. Once more, the two conditions have been quite distinct and have even been antagonistic; and lastly, they have been identical and with no

---

* " Handbuch der rationellen Pathologie," 1846-53.
† " Researches and Observations on the Causes of Scrofulous Diseases." Translated by W. H. Ranking. London, 1844.
‡ "On the Nature of Scrofula," by Richard Carmichael. London, 1810, p. 21.
§ " Observations on Scrofulous Affections," by R. Hamilton, M. D. London, 1791, p. 65.

line of separation between them. The very term "tubercle" has experienced a violent series of fluctuations. It has been applied first to one appearance and then to another; its limits have been terribly curtailed; vaunted specific features have one by one been removed from it, until it must be owned that the tubercle of to-day is but a poor and bald affair as compared with the tubercle of the time of Laennec.

The forced association of scrofula. therefore, with this vague pathological element, the very Proteus of pathology, can in some way account for the uncertain position the disease has occupied from time to time, and for the somewhat indefinite outlines it still retains. At the same time it must be owned that much good has been done by the extensive investigations that have been pursued on the subject of tubercle, although they have tended a little too much towards an attempt to demonstrate a specific element for a certain class of disease. Thus it happens that few additions to our knowledge of the histology of so-called tubercular affections have been made that have not been seriously burdened by theoretical matter. New facts have been immediately hampered with some new theory or some old dogma, and the results of pure investigation have often been lessened in value by speculation and conjecture. These difficulties, clinical and pathological, while they can perhaps explain the vagueness that still marks the scrofulous process, may indicate the direction it would be well to pursue in any subsequent inquiries.

CHAPTER II.

SCROFULA AND TUBERCLE.

Inasmuch as scrofula is so closely bound up with the subject of tuberculosis, the first point to be considered in discussing the pathology of the former disease is the nature of tubercle and its relations to the scrofulous process. It is obvious that no definition of scrofula can be attempted until this relationship has been clearly set forth.

The term "tubercle" was originally applied to a certain naked eye appearance, to little distinct specks or spots of diseased tissue that were conspicuous as nodules or tubercles. When first used the term "tubercles" had no more clinical significance than has the term "nodular;" and it is remarkable to note how in time a clinical meaning of the most emphatic character attached itself to the term, and has since clung to it. So close is this connection that it is almost impossible to separate certain anatomical appearances from certain clinical conditions; and no matter to what structural change the word tubercle is, or has been, applied, there still lurks behind it a subtle suggestion of a distinct clinical state, known as the tubercular condition. The term was for a time applied to many states of tissue, that although anatomically different, yet possessed the common feature of being nodular in outline. A better restriction of the word was arrived at when it was set forth that some of these nodules were grey and clear, while others were yellow and opaque. Thus arose a division of tubercle into the grey and the yellow varieties. The yellow, or so-called crude, tubercles were for the most part caseous masses, or at least masses advanced in that decay; and they had soon to be eliminated from the domain of tubercle when it was shown that caseation was by no means limited to what was known as the tubercular process. With regard to the other tubercles, the grey variety, it was found that such nodules when met with in the lung were often made up solely of little masses of alveolar epithelium, the results of a lobular catarrh. All such nodules, therefore, had to be eliminated, and still finer distinctions laid down as characteristic of tubercle. The term was then restricted to such grey semi-transparent bodies as were not merely masses of catarrhal exudation, and that, while retaining the size of a millet seed, were hard and firm. These tubercles, it was noted, in time became opaque in the centre and then wholly caseous, and had a tendency to fuse together and form larger masses. The name of miliary tubercle was given to them, and in the disease known as acute miliary tuberculosis they were considered to be met with in perfection. In time, however, certain tissue changes were noted, which were regarded as tubercular, but which were not associated with the appearance of these distinct grey masses. In the place of such masses certain microscopic nodules alone were detected that were found to possess a fairly simple structure; and as it was observed that certain of the grey miliary tubercles, visible to the naked eye, were simply made up of a collection of these microscopic nodules, the latter were distinguished by the term submiliary tubercle. It must be owned that for a long while the microscopic features of tubercle were very indefinite and confused, and it was not until this finer restriction of the word was adopted that anything like uniformity was obtained in histological descriptions. These little microscopic nodules were found to be of common occurrence, and capable of undergoing the final degenerative process, without having first formed themselves into the larger masses known as miliary tubercles. It is to these microscopic nodules only that the bare term "tubercle" is, in its strictest sense, now applied. Thus it will be seen that the anatomical ground on which tubercle rests has been from time to time curtailed, and that the large basis it originally possessed has been cut down at last to a very minute point. This submiliary mass, this ultimate tubercle has been described by many observers, and has received many names; but although the terms used differ, and although some descriptions are a trifle modified from the rest, yet there is so much general accord that pathologists of the present day appear to be at least agreed as to what tubercle looks like, even if they disagree as to what it is. It must be understood that tubercle in its simplest sense refers to the most typical stage of a certain tissue change, and that to the process that preceeds its appearance, as well as to that that follows, the term tubercular can be applied.

Histology of Tubercle. — This simple submiliary

tubercle has been described under many names, as "primitive or elementary tubercle" by Köster, as "tubercular follicle" by Charcot, as "reticular tubercle" by E. Wagner. All these terms may be regarded as synonymous. The structure of such a tubercle is this:

It is composed of a mass having a fairly rounded outline, and made up principally of cells. These cells are so arranged as to form in typical specimens three zones. The central part is occupied by one or more giant cells, round this is a zone of many so-called epithelioid cells, and beyond this is a third zone of simple embryonic cells or leucocytes. All these cell elements are supported by a fine reticulum, which is generally concentrically arranged at the periphery, and towards the centre is observed to be continuous with the processes that commonly come off from the giant cells. The affected district is non-vascular. Such is a typical tubercle. Modifications of structure are, however, permitted. The giant cell may occupy the periphery, or may be entirely absent, in which case the appearance is supposed to be maintained by the character of the cells in the mass, their obvious changes, and their general arrangement. The giant cell, although not specific of tubercle, is usually present, and if any differentiation is adopted with regard to terms, the term "follicular or reticular tubercle" would apply to the perfectly developed mass, the terms "elementary or primitive tubercle" to those nodues that show less perfect evolution, and possess perhaps no giant cell.* As to the structural origin of tubercle it will be easily understood that great diversity of opinion exists. Some hold that it is developed from connective tissue, and is a connective tissue growth;† others, that it is essentially a lymphoid or adenoid structure.‡ Some observers—and among them Cornil and Ranvier§—refer its origin to

---

* For a brief but excellent account of the grades of tubercle, see Art. by Dr. Grancher in "L'Union Medicale," vol. xxxi. 1881, p. 873.

† See exposition of this view, by Dr. D. J. Hamilton. "Practitioner," August 1881.

‡ "On the Artificial Production of Tubercle," by Dr. Wilson Fox. London, 1868, p. 25,

§ "Manuel d'Histologie pathologique. Paris 1881, vol. i. p. 236.

the vessels of the part, and state that a coagulum forms in the blood capillary, the endothelium of whose wall vigorously proliferates, so that on section the coagulum is seen to form the mass of a giant cell, and the proliferated endothelium its many nuclei. If the vessels be larger changes take place in its walls, and the various zones of the tubercle are then considered to correspond to the various tunics of the artery.* Others, again, regard the giant cell as a protoplasmic mass, and consider that it indicates a return of the tissue to a more embryonic state.† Those who hold this view believe that all gradations can be observed between the epithelioid cells and the leucocytes on the one hand, and the giant cells on the other. Lastly, among other views may be noted one that applies only to the lung, and is to the effect that these giant cells are formed by the fusion of the epithelial cells of the lung alveoli.‡ These —although but a few of the theories that have been advanced—are perhaps the most representative, and will be more fully discussed in the chapter on the Pathology of Scrofulous Glands.

Such being tubercle, the first question to be asked is this—Does this tubercle present any specific anatomical element? It assuredly does not. Lebert§ some years ago endeavored to establish the specific character of certain cells in tubercle, the so-called "tubercle corpuscles," but his conclusions were soon found to be erroneous; and indeed these "corpuscles" were none other than shrivelled cells, not distinguishable from shrivelled pus corpuscles. Schüppel,‖ again, in more recent times endeavored to maintain the specific character of the giant cell, and urged that this structure was peculiar to tubercle, and indeed diagnostic of it. This argument

---

* M. Kiener. "Discussion before La Societe Medicale des hopitaux. L'Union Medicale," vol. xxxi. 1881, p. 316.
† See on this point "Epeer. Untersuch. uber die Herkunft der Tuberkelelemente," etc. Wurburg, 1875, by E. Ziegler.
‡ Dr. E. Klein. "The Anatomy of the Lymphatic System—The Lung." London, 1875, p. 76.
§ "Physiologie Pathologique. Paris, 1845.
‖ Dr. Oskar Schuppel. "Untersuchungen uber Lymphdrusen Tuberkulose. Tubingen, 1871.

has, however, been overthrown, and it is now known that giant cells are to be met with under the most varied circumstances and in conditions that could in no way be termed tubercular. Thus they have been found in simple chronically inflamed connective tissue, in chronic ulcers, in gummata, in erosions of the os uteri,* and in many other parts and tissues. Although giant cells cannot be regarded as special to tubercle, yet it must be owned that they are not commonly met with unassociated with that product. The anatomical individuality of tubercle therefore depends upon no especial factor or element, but must rest upon the general conformation of the mass, the grouping of its parts, the relation it holds to the tissues around, and above all to its history, its tendencies, its peculiar progress. The most remarkable features of the nodule are perhaps its early non-vascularity, its tendency to caseation, its aptitude for spreading locally and, under certain conditions, generally, and its action when inoculated experimentally in animals. Other points in the history of tubercle will be considered subsequently.

Scrofula and Tubercle.—The ground is now prepared for a discussion of the relations between scrofula and tubercle. As a preliminary step some definite meaning must, for the purpose of this discussion, be attached to the terms "scrofulous" and "tuberculous." It is obvious that at the present stage of the enquiry those terms cannot be used in an anatomical sense, and must therefore be applied temporarily to certain clinical conditions. For the present purpose then the terms "tuberculosis" and "tuberculous" will be considered as applying to such diseases as acute miliary tuberculosis, tubercular peritonitis, tubercular meningitis, and the term "scrofulous" to those diseases commonly known by that name, as, for example, glandular enlargements, certain chronic bone and joint affections, cold abscess, certain ulcers and eruptions of the skin and mucuous membranes.

It would be well to omit phthisis from either cate-

---

* Dr. Carl Friedlander. "Ueber locale Tuberculose." Volkmann's "Sammung." No. 64, 1873.

gory, as it will be separately discussed by-and bye. Is tubercle—as just described—met with in scrofulous affections? To this question an affirmative answer must most certainly be given. In scrofulous lymphatic glands the most perfect and most typical tubercle is to be met with; in the synovial membrane, and in the bone in cases of so-called strumous joint disease perfect tubercle has been discovered. Dr. Lannelongue has lately shown the origin of cold abscess from tubercular deposit, and demonstrated the progress of such abscesses by extension of the so-called tuberculous process. Tubercle also is to be found in the floor of scrofulous ulcers, in lupus, in certain affections of the mucous membrane, and in other parts. If this be the case there would appear to be no difficulty in establishing the fact that scrofula is what is termed anatomically a tubercular process. But this proposition is not so simple as it seems, and much of the difficulty that surrounds it is due to the severe and definite clinical meaning that attaches to the term proposed.

In the first place, these tubercles are not met with in all scrofulous affections. In the superficial skin eruptions, in some of the more common affections of the mucous membranes, and in many typically scrofulous glands, no tubercle is to be met with. And, again, the presence of tubercle has not been demonstrated in all cases of disease of bone and joint observed in the scrofulous. Thus it happens that some pathologists would limit the term "scrofulous" to those affections only that present no tubercle, and reserve for the rest the term "tuberculous." It is chiefly with regard to the lymphatic glands that this division of disease has been urged, and by no one more vigorously than by Cornil.* But it must be remembered that there are grades and degrees in the tubercular process, just as there are varieties and degrees of inflammatory action. In some cases the tubercular action—if I may be allowed the term—does not proceed so far as the formation of tubercle, just as all inflammations do not

---

* *Journal de l'anatomie et de la physiologie* (Charles Robin). Paris, 1878, No. 3.

always proceed to the formation of pus. At any point in the tubercle-producing process the action may end and caseation set in.

Many scrofulous glands caseate without developing any tubercle, but in the process that precedes such caseation one recognizes a state that is at least preliminary to the formation of the nodule, a pre-tubercular state as it may be termed. Tubercle is the most finished structural change of a certain process, and such a *période d'état* may never be reached in a vast number of strumous disorders. Now it appears unreasonable to make a conspicuous division between these two grades of gland affection—the gland that shows tubercle and the gland that just falls short of that product. They are both essentially tubercular, and terms should not be applied to their morbid conditions that would indicate more than differences in degree. But those who assert the distinction of scrofula from tuberculosis term those glands that show perfect tubercle tuberculous, and those that present only the immature structure scrofulous. This use of terms would not be objectionable had it only an anatomical basis, and did not a very rigorous clinical meaning associate itself with these two adjectives. Cornil, and those who follow his teaching, do not limit themselves to structural differences, but lay down clear clinical distinctions between the scrofulous and the tubercular gland. As regards at least the external glands—and it is with these that we are now most concerned—I must assert that these clinical distinctions are not clearly marked and can hardly be maintained. I own that there are affections of internal glands (those of the mesentery, for example) that present perfect tubercle, and yet follow a course so distinctive and often so detrimental that they are clinically removed from the category of what is commonly known as scrofula. Such glands, however, differ from the simply scrofulous—to use that term in its usual sense—only in degree, and the evils that attend their development depends mainly upon the cause and locality of the disease.

It will be said that all this is merely a matter of

terms. It is so; but certain circumstances conspire to make it also a matter of importance. The term "tubercular" is used in a double sense. It is applied to an anatomical condition—to any disease presenting perfect tubercle—and it is also applied to certain clinical states. Unfortunately these two conditions do not quite coincide and the presence of tubercle does not, of necessity, imply that grave state of health associated with the word "tuberculosis." Tubercle has to a great degree been discussed in connection with acute miliary tuberculosis and certain lethal lung affectioms; and from ancient custom or ancient bias it has somehow become associated with all the grave clinical import of these diseases. This is much to be regretted, and constitutes a bias of evil consequence. In a little patch of lupus on the face perfect tubercle may be found. Is a patient so affected to be called tuberculous in the usual clinical sense? Is he likely to die of some acute and sudden tubercular mischief? Is he not, on the contrary, as likely to attain old age as is the majority of other persons? One can quite understand from a case such as this the vigor with which the application of the term tuberculous to such a disease would be combated. Perfect tubercle is met with in lymphatic glands, but such glands after a time may eliminate the disease and a cure result followed by no bad consequence. Leaving even scrofulous affections, one finds tubercle in other diseases quite remote from any clinical association with tuberculosis. Köster has described miliary tubercle in osteo-myelitis, in chronic pericarditis, in the primary syphilitic sore, in elephantiasis of the labia, and in other conditions. To apply the clinical term "tuberculous" to such cases would simply be ridiculous. This injurious use of the term proceeds, I imagine, to a great degree from the general attempt to associate every diathesis and disease with some specific anatomical element. The logic has been as follows: Tubercles are found in acute miliary tuberculosis and other fatal diseases; these diseases are called tuberculous, therefore every other disease that presents tubercle must also be tuberculous.

As M. Ferrand * well observes, "tubercle does not constitute a disease any more than does suppuration." It is the exclusive indication of no one malady and the outcome of no one special state of defective health. Although one recognizes the fact that tubercle appears in scrofula, yet one is positively loth to term scrofula a tuberculous disease; and I would almost go so far as to say that it would be well not to call it a tuberculous disease, until the bias associated with the latter term has been removed, until that term is accepted in a more generous and rational sense, and until it ceases to attempt to force an alliance between an anatomical appearance and a clinical state.

To return to the relationship between scrofula and those affections that may still for distinction be termed tuberculous. Dr. Grancher † urges that in scrofula an immature or embryonic tubercle is alone met with, and that the adult or completely developed tubercle does not usually occur in the disease, but, on the contrary, is the main attribute of tuberculosis exclusively so called. He regards scrofula therefore as "une tuberculose atténuée," as a milder, less perfect, less developed form of tuberculosis. He owns the perfect identity of the two affections, and insists that they differ only in degree, in age, in the matter of maturity. In this sense, therefore, he speaks of scrofula also as "une tuberculose naissante," "une tuberculose au premier degré." Here, however, it must be stated that the immature tubercle of Grancher corresponds fairly to the tubercle already described as the elementary or primitive tubercle, or tubercular follicle; while the adult tubercle of which Grancher speaks is represented by the grey granulation of Laennec. This mass, often visible to the naked eye, is well known as the simple miliary tubercle. It is hard and firm, has a tendency to fibrous transformation, and is merely a conglomeration of the smaller and simpler tubercle masses already

---

* *L'Union Médicale*, vol. xxxi. 1881, p. 40.
† "Dictionnaire encyclopedique des Sciences Medicales. Art. Scrofule." Paris, 1880 p. 304.

described. This conglomeration is fairly considered to indicate a more complete development of the process.

It is certainly true that this large adult granulation is very rarely, if ever, met with in truly scrofulous affections. In scrofula the tubercular process seldom attains to so elaborate a structure, but ends in caseous degeneration before the formation of such a mass is reached. Dr. Grancher, moreover, takes a wide view of the structure of this embryonic or immature tubercle found in scrofula. He asserts that although it often attains to that perfection of structure we have ascribed to tubercle it might still be represented by a much less definite tissue. It may appear as little more than a cluster of embryonic cells recognizable as on the whole tubercular by its arrangement, its progress, its general tendency, and termination, as well as by its surroundings. Such a tissue would correspond to Virchow's "granulation tissue." In more fully developed masses an arrangement of cells and giant cells would be observed more approaching that of the perfect tubercle, and would fairly accord with the description of primitive tubercle as given by Köster. To this "immature tubercle" which, I must repeat, includes not only the tubercle as we have defined it, but also those less definite masses called by Virchow "granulation tissue" and by Cornil "îlots strumeaux," he proposes to give the name of "scrofulome."

Other relations between scrofula and tubercle have been maintained. Rindfleisch[*] considers scrofula as the starting-point of the tubercle-producing process. He considers that the tubercle is usually derived by a process of infection from some near or distant seat of scrofulous disease. Thus he traces the connection between certain eruptions of tubercle in the lung and a preliminary scrofulous bronchitis. Such being the connection existing he would regard scrofulous patients as extremely liable to those serious clinical conditions distinctly termed tuberculous. He states that tubercu-

---

[*] "Chronic and Acute Tuberculosis. Ziemssen's *Encyclopædia of Medicine*," vol. v. p. 638.

losis (in a clinical as well as anatomical sense) seldom
oocurs except in such patients. If tubercle appears in
a gland then has it been due to a previous scrofulous
trouble, to a scrofulous catarrh of a mucous surface, or
some similar lesion. Or, to take another example, a
tubercular ulcer may commence as a catarrh, and after
a time tubercles will appear at its base. Here he
regards the preliminary catarrh as scrofulous, the subsequent local eruption of tubercles as tuberculosis.
Such a distinction as this is cumbrous, and appears like
pedantic elaboration. It is evident, moreover, that
Rindfleisch limits tubercle by clinical lines, and restricts
it to those more fatal diseases associated with the
appearance of tubercle. His distinctions indeed appear
to be almost wholly a matter of terms, since he does
not deny thé very close and indeed direct connection
between what he, however, maintains to be two distinct
morbid states.

Many other views as to the connection between scrofula and tuberculosis depend upon the results of experimental inoculation. Some pathologists, for example,
regard tuberculosis as an infective disease, a disease not
so much due to hereditary diathesis as an acquired
malady like syphilis or glanders. Relying upon the
well-known inoculation experiments they urge that a
true diathesis cannot be transmitted by inoculation,
and that as tuberculosis may arise from inoculation it is
therefore not a diathesis; whereas the products of scrofula not being inoculable that malady may be ranked
with the diatheses. As to the relations between the
two they would urge that they are the relations of soil
to seed. Scrofula is the soil, tubercle the seed, and the
point of contact between the two conditions is this—
that it is especially, if not exclusively, upon the soil of
scrofula that the infective tubercle can take root and
develop. M. Rendu † has vigorously espoused these
views in a recent communication, and his conclusions
may be considered as representative of a large class of
theories. The introduction into the question of an
infective character for tubercle brings us to discuss:

---

†*Scrofule et Tuberculose.* *L'Union Medicale*, vol. xxxi. 1881, p. 53.

The Experimental Inoculation of Tubercular and Scrofulous Products.

These experiments have been very elaborately conducted by Villemin, Burdon Sanderson, Wilson Fox, Klein, and others, and in more recent times by Cohnheim, Hueter, Schüller, Klebs, and Deutschmann. The earlier experiments consisted in inoculating "certain animals" with tuberculous matter. The material in these experiments was injected into the pleural or peritoneal cavities, or introduced under the skin. The result was that in most cases the animals operated upon developed a disease considered as akin to acute miliary tuberculosis in man. The real nature of this produced disease and its mode of development are open to considerable dispute, but as in a subsequent chapter I propose to discuss the value of these experiments, I will, in this place, merely refer to them as means of diagnosing between scrofula and tuberculosis. It was argued from these early experiments that tuberculosis was an infective disease, and that this feature distinguished it from scrofula. But then the result of certain investigations—such as those by Wilson Fox—showed that when this tubercular matter was introduced under the skin it set up a kind of local scrofula, a suppurative process associated with enlarged and subsequently caseous glands. This then at once appeared to show that the relation between scrofula and tuberculosis was very close, and a matter of difference in degree rather than in kind. The only way out of the difficulty was to call the local manifestations tuberculous. This was, of course, done. But then it was found that scrofulous matters, if used for these experiments, produced the so-called general tuberculosis as readily as tubercular matter. Indeed the matter from a caseous gland became the most active and favorite agent in these experiments. Those who still clung to the belief that tubercle had the distinguishing feature of being inoculable claimed these cheesy glands as tubercular, and so the field of scrofula was narrowed by the loss of its most typical manifestation. More recent experiments, however, now show that portions of the fungous granulations from "white swellings" of joints, the pus from cold abscess, granulations from so-called scrofulous osteitis and periostitis can all produce general tuberculosis when inoculated. Cohnheim,* whose experiments are most elaborate and extensive, asserts as the result of his researches that all the recognized tubercular and scrofulous processes, however different anatomically, are tubercular, inasmuch as the products of all of them are almost equally active on inoculation. So far, then, the identity of the two affections appears to be confirmed by these experiments; and as it would seem that products from strictly tuberculous diseases often produce readier results than those from strictly scrofulous diseases, it may be further argued that the two morbid conditions differ only in degree as has been elsewhere maintained. But even now some pathologists still advance the opinion (which I give in the words of M. Villemin) that "tubercle alone gives tubercle by inoculation." Those who retain this view are compelled to exclude from scrofula all its classical features; the caseous gland, the cold abscess, osteitis, periostitis, and "white swellings;" and all that they can leave for the disease are a few superficial lesions, the products of which will in time be probably found to be inoculable, and then for M. Villemin and his followers scrofula will be an extinct disease. It is needless to discuss this point further. It is obvious that rigid differences between scrofula and tuberculosis (to keep to the old divisions) cannot be founded upon experiments. The discussion becomes again merely a matter of terms, and a conflict between clinical and anatomical standpoints. So far, however, as these experiments affect the present subject they show:—1. That tubercular matter when introduced into the bodies of certain animals can produce at first a local disease not distinguishable from scrofula. In addition to the results already mentioned, M. Kiener†

---

* " Die Tuberculose vom Standpunkt der Infectionslehre." Leipsig, 1880.
† *L'Union Medicale*, vol. xxxi. 1881, p. 316.

has shown that the injection of tubercular matter into the testis can induce caseous inflammation of that body, and into the knee-joint, a chronic joint disease that fully accords with the common notions of white swellings. Cohnheim's experiments have all the same bearing, although these observers may refrain from applying the term scrofulous to the results produced. 2. That scrofulous matter when used as a vehicle for inoculation can produce general tuberculosis. 3. That tubercular matter acts often more vigorously in these experiments than does strictly scrofulous matter. From these results it may be gathered that experimental inoculation maintains the identity of scrofula with tuberculosis, and at the most can only show that the two conditions differ somewhat in intensity or degree.

I must again point out that throughout this discussion the terms "scrofulous" and "tuberculous" or "tuberculosis" are used in the sense detailed at the commencement of this chapter, and refer to certain definite clinical diseases, and not to the anatomical bases of any disease.

The following conclusions may be stated as to the relations between Scrofula and Tubercle.

1. The manifestations of scrofula are commonly associated with the appearance of tubercle; or if no fully formed tubercle be met with, a condition of tissue obtains that is recognized as being preliminary to tubercle. Anatomically, therefore, scrofula may be regarded as a tuberculous or tubercle-forming process.

2. The form of tubercle met with in scrofulous diseases is usually of an elementary and often of an immature character. Whereas in diseases called tuberculous in a strict clinical sense, a more perfect form of tubercle is met with in the form of the grey granulation or "adult tubercle" (Grancher).

3. Scrofula, therefore, indicates a milder form or stage or tuberculosis, and the two processes are simply separated from one another by degree.

CHAPTER III.

THE NATURE OF TUBERCLE.

As scrofula is a disease associated anatomically with the appearance of tubercle, the pathology of the affection naturally rests to a great extent upon the nature or morbid significance of this remarkable structure. As may be supposed, a vast number of theories have been advanced in connection with this matter.

To discuss even the most prominent of them would be beyond the scope of this book, and I will content myself, therefore, by detailing what appears to me to be the most reasonable explanation of the nature of tuberculosis. I believe that the view at present most generally accepted with regard to tubercle is that it is a neoplasm or new growth; a connective-tissue growth according to some, a growth from adenoid or lymphoid tissue according to others. Whether this neoplasm is of embolic origin or whether it is developed *in situ*, is of no concern to the present subject. The fact remains that in spite of possible differences of origin, a number of pathologists regard tubercle as a new growth. I, however, would venture to urge that tubercle is merely the product of a peculiar form of inflammation; that it is no neoplasm in any other sense than that it is an inflammatory neoplasm. The inflammation with which it is associated has many and distinctive features, and these, when even fairly marked, can separate it from every other phase of inflammatory action. What are these peculiarities need not here be discussed. They will be fully dealt with subsequently. The only point to establish now is whether the mass tubercle is, or is not, a direct product of inflammation. On this point, I will draw attention to these facts. The appearance of tubercle is frequently preceded by an inflammation of undoubted character elsewhere, which stands to the nodule in the relation of cause to effect. As a conspicuous instance of this, I might cite certain gland enlargements. An enlarged cervical gland can often be definitely traced to some perfectly simple inflammation, seated, let us say, within the mouth. Now one knows the well marked tendency of lymphatic glands to accurately reproduce morbid conditions transmitted to them from the periphery, and I would urge that the present instance forms no exception to that rule. The affected gland shows at once evidences of inflammation, and, in fact, faithfully reproduces the process active at the periphery. But in time the simple action assumes more peculiar features: the products of the inflammatory process become themselves peculiar; they mass themselves together in a strange manner, and conspicuous among those products is tubercle. At no time in the course of the gland affection could one say, here inflammatory action ends and the growth of a neoplasm begins. In no case of gland disease that I have yet met with has the process commenced by the deposit of tubercle in a tissue that is absolutely unchanged from its normal condition; and although at least one pathologist * speaks of such an occurrence, I must venture to doubt its reality until it is supported by more detailed evidence. As will be seen in the chapter which deals with the pathology of the gland affections, the appearance of tubercle is always preceded by changes distinctly inflammatory; changes marked by increased vascularity of the part, extensive exudations and active cell proliferation; and, although as the sequel shows, this inflammation assumes distinctive features, yet its general nature remains unaltered. Rindfleisch † says that "it is often impossible, in a given tubercular lesion, to determine how much is inflammatory and how much tubercular." Dr. Lannelongue,‡ who is a firm believer in the entity of tubercle as a neoplasm, is yet constrained to observe in dealing with tuberculosis of bone, that the osteitis with which it is associated very often makes its appearance before the

---

* "Pathology and morbid anatomy," by Dr. T. H. Green, 4th ed. 1878. Fig. 70, p. 241.
† "Ziemssen's Encyclopædia," *loc. sit.*, p. 647.
‡ "Absces Froids et Tuberculose Osseuse." Paris, 1881, p. 133.

tubercle is met with. M. Keiner,* speaking of tubercles of serous membranes, states that these structures at first differ in no way from the products of simple inflammation. Other observers maintain the same fact, and yet when the products become peculiar, inflammation is considered as withdrawn from the field. If it is allowed that the process is at its outset inflammatory, it seems most illogical to assume a gross and sudden change of pathological action, simply to explain appearances that happen to differ from those met with in more familiar inflammations. Presuming, on the other hand, tubercle to be a neoplasm, it is remarkable that throughout its whole course it should be so very frequently and intimately associated with inflammatory change. No other neoplasm shows this extraordinary alliance. I would venture to say that no new growth with which pathology has rendered us familiar makes its *début* so often associated with inflammatory change, and its ending so often marked by suppuration and death of tissue, as does this structure tubercle. It may be that the reputed neoplasm excites inflammation in the adjacent structures; but if that be allowed, it is strange that it should also be preceded by inflammation, and stranger still that a peculiar form of irritative inflammation should be considered by many as needful to procure this neoplasm. Then, again, it is important to note the elevation of temperature that is associated with the appearance of tubercle, and that would point also to its inflammatory nature. Dr. Thaon, who supports the theory that tuberculosis is due to a special inflammation, remarks that, "like all inflammations, it is always accompanied by fever, a circumstance not observed in cancer." †

Lannelongue has also drawn attention to the local and general rise of temperature noticed in cold abscess, an affection that he has shown to be of tubercular nature, both in its origin and also in its subsequent

---

* " De la Tuberculose et des Affections dites scrofuleuses." *L' Union Medicale*, vol. xxxi, p. 316, 1881.
† " Recherches sur la Tuberculose et la Scrofulose." *L' Union Medicale*, vol. xxxi. 1881, p. 41.

progress. M. Du Castel,* a supporter of the theory as to the inflammatory nature of tubercle, also refers to Peter's observation on the local elevation of temperature at the commencement of phthisis as pertinent to the present subject.

Then, again, the curability of many of these tubercular affections, especially those that are more properly included in scrofula, would appear to support the theory of an inflammatory nature, and to militate against the idea of a neoplasm, particularly as that neoplasm shows evidence of being by no means of an innocent nature.

A neoplasm that can multiply as fast as tubercle appears sometimes to multiply, that can spread with so marked a determination, that can invade and occupy an entire organ, must be regarded as a growth from which a spontaneous cure is hardly to be expected. Yet perfect tubercle may occupy the the length and breadth of a gland, and yet that gland leisurely caseates, suppurates, and heals. Moreover, the tendency sometimes shown by tubercle to form a fibrous material resembles in a striking manner the tendency commonly exhibited by the granulations in a simple wound to develop connective tissue; and such a transformation—although it may be only occasional in tubercle—does not accord with one's notions of an active new growth. Very strong support has been given to the theory of the inflammatory nature of tubercle by the ingenious experiments of Ziegler.† Ziegler induced inflammatory changes in dogs and rabbits by inserting under the skin or in some of the cavities of the body two thin discs of glass so cemented together that fine interstices were left between them. There could be nothing specific about two little discs of glass, and the animals experimented upon were in sound health. The discs were removed at varying periods and examined. First the interstices between the pieces of glass were found to be occupied by a mass of leucocytes. These

---

* *L'Union Médicale*, vol. xxxi, 1881, p. 138.
† "Experimentelle Untersuchungen uber die Herkunft der Tuberkelelemente," &c. Würzburg, 1875.

often underwent degenerative changes, and showed no tendency to form any definite structure. In other instances vessels were observed, giant cells, and so-called epithelioid elements made their appearance (developed, Ziegler presumes, from the leucocytes), a reticular tissue formed, and a tubercle was the result. Here the inflammatory nature of tubercle can hardly be called in question, and the only objection that in this instance can be raised against such a theory lies in the bare statement that in the class of animals operated upon there is a tendency for the growth of the neoplasm tubercle to be excited by slight and non-specific irritation.

Birch-Hirschfeld,* referring to these experiments, and arguing also upon more extended bases, asserts that "tubercle might be regarded as a degenerated species of inflammatory neoplasm (granulation) determined by necrobiotic processes." This definition affects certain peculiarities of the tubercular process, with which at this moment we have no concern, but in the general principle it embodies it may be entirely accepted.

The peculiar form of the inflammatory process that leads to the appearance of tubercle appears to have no remote parallel in the inflammatory changes peculiar to tertiary syphilis. A trifling wound or abrasion inflicted upon a patient suffering from tertiary syphilis often takes on a very remarkable action. A similar lesion in a healthy individual would probably heal without perceptible local disturbance. Owing to a previous blood disease the tertiary syphilitic is liable to an inflammation of a perfectly distinctive type; so distinctive that it is known as gummatous and its specific product as a gumma. The microscopic appearances of a gumma do not happen to be of so remarkable a character as are those incident to tubercle, but none the less is a gumma a distinct and marked deviation from the ordinary type of inflammatory product. No one would, I presume, assert that a gumma is a neoplasm in any other sense

---

*Ziemssen's " Cyclopædia of Practical Medicine, Art. Scrofulosis," vol. xvi. 1877, p. 758.

than that it is an inflammatory neoplasm, and from the standpoint of general pathology I fail to see why a like character is not allowed to tubercle. In both tuberculosis and in tertiary syphilis there is a tendency to develop inflammation of a peculiar type, in the one instance that peculiarity shows itself by the product tubercle, in the other by the product gumma.

I will make no mention in this place of the tubercles in acute miliary tuberculosis. That disease has certain traits that are peculiar to itself, and that remove it from general consideration. To introduce the subject into the present question would be akin to thrusting pyæmia into a consideration of the pathology of simple inflammation.

CHAPTER IV.

THE INOCULABILITY OF TUBERCLE.

The experiments to demonstrate the inoculability of tubercle have been very extensive, but on the whole somewhat meagre of good results. Caseous matter from a tubercular source injected into the pleural cavity of rabbits, pigs, and other animals, induced a fatal disease associated with an eruption of so-called miliary tnbercles in the lungs and other parts. This disease was considered as akin to acute miliary tuberculosis in man. From experiments of this character tubercle was considered to be due to infection, to be the outcome of some occult virus, and its pathological position to be among infective disorders. There is no doubt that these injections of caseous matter induce a fatal disease associated with the eruption of little tubercular nodules, but the objections raised to the conclusions and theories resulting from these experiments are somewhat obtrusive. In the first place, is the disease induced in these animals acute miliary tuberculosis? Many urge that it is not, but that it is rather of the nature of pyæmic infection. Wagner indeed failed, by these experiments to develop a disease that he could regard as tubercular. It appeared to him rather to be a chronic pyæmia. Then, again, the majority of the so-called miliary tubercles were found to be merely masses of adenoid tissue developed from the lymphatic structures of the affected part ; and it has very reasonably been urged by Friedländer that as these masses show neither epitheloid cells nor giant cells they can hardly be classed as true tubercles. In justice, however, to those who maintain the tubercular nature of these lesions, it must be stated that in the lungs of animals with this disease undoubted tnbercular masses are met with, although they may not be the predominant feature. Certain experiments of Dr. Wilson Fox appear to raise a more serious objection to the infection theory. He found that the injection of caseous matter was not essential for the production of this artificial tuberculosis. It was sufficient in certain animals ot induce suppurative inflammation by means of a setou simply passed beneath the skin ; this inflammation became caseous, and general tuberculosis followed. Like experiments by others have verified these results, and in connection with such may be mentioned the injection of minute non-animal irritants. Here, then, it would appear that the animal had the power of manufacturing the infecting material, that the virus could easily be generated *de novo* ; and it must be owned that this supposition somewhat militated against those notions of infective disease that are derived from a study of syphilis as a type of such disorders. But in connection with this point it was asserted that certain animals had a great tendency to develop caseous inflammation from trifling causes, and as it was known that caseous matter is not the only means of conveying the infection, it must be persumed that the specific virus was developed at the same time.

Buhl * asserted, *à propos* of the same matter, that in all cases of general tuberculosis in man, a previous case-

---

* " Lungen-Entzundung, Tuberculose, und Schroindsucht." Munchen, 1872.

ous mass could somewhere be found, from which mass the body had been, as it were, infected. Rindfleisch,* going a step further in the same direction, is inclined to maintain that even local tuberculosis proceeds from some previous caseous inflammation, as an inficting focus. Such inflammations he considers to be scrofulous, and hence the close connection he traces between scrofula and tnberculosis as between a cause and an effect. Quite recently Dr. Creighton † has given reasons for opposing the views of Buhl and others, and concludes that disseminated tuberculosis does not originate in a primary source of infection within the body, but that the infecting agent is a virus introduced into the body from without.

The whole question therefore is still unsettled. In all the experiments above alluded to it was noted that local changes occurred at the inoculated spot, that these changes extended locally within certain limits, and then became general. In Dr. Klein's ‡ experiments, for example, where matter was injected into the pleural cavity, a chronic pleuritis was the first result, the morbid changes in the pleura could then be traced by direct continuity of tissue into the lungs, where they soon became more general. In like manner, in the researches of others it was shown that matter injected into the peritoneal cavity § first induced tuberculosis of the peritoneum, then of the mesenteric glands, then of the pleura and mediastinal glands, and lastly of the lungs and other viscera. So with regard to the connective tissue the same direct continuity of diseased action was observed. Caseous inflammation in that tissue was followed by changes in the corresponding glands, from whence the process spread more generally. M. Kiener has shown that if the matter was injected into the knee or the testis, local tubercular change took place in those parts before any general infection occurred. and that

---

* *Loc. cit*
† International Medical Congress, 1881. " Abstracts of Papers," sec. iii. p. 39.
‡ " Anatomy of the Lymphatic System,—The Lung," 1875 p. 47.*et seq*.
§ See note of these experiments, by M. Kiener, *loc. cit.*, p. 349.

when such infection did occur a structural continuity in the morbid processess could be detected. It must therefore be allowed that the tubercular process when once set up has a remarkable tendency to spread, and that such spreading appears to be mainly promoted by the lymphatics; that it is a process that infects locally must be admitted, but that it constitutes a disease peculiar in so far that it can be transmitted unmodified from one individual to another would appear to require the support of some further facts. It must be freely acknowledged that in certain animals the inoculation of tubercular matter produces a disease having an eruption of "tubercles" as its principal anatomical features; but to maintain the theory as to the infective nature of tubercle it must be shown that no other matter or tissue can produce a like effect, that the results obtained are independent of any peculiar morbid tendencies possessed by the animals experimented upon, and that—while a reasonable margin is allowed—the result of such inoculations are constant. Many clinical facts support the idea that tubercular disease may be transmitted from one individual to another.* With regard to animals this transmission has been chiefly insisted on in cases of pearl disease (perlsucht) in cows, it being stated that pigs when fed upon the milk from animals so affected become in many instances themselves tubercular.† Dr. Creighton ‡ has lately urged that this disease can also be communicated to man, the vehicle being the milk or flesh of the affected animal. With regard to man also, so many carefully considered cases have been recorded à propos of the communicability of phthisis that it is hard to resist the conclusions urged by those who advance them. An excellent summary of

---

* See, for example, a case by Dr. Guerin ("Discussion sur la Tuberculose." Bull. de l' Acad. 1867,) where a man with tuberculosis infected his wife. He died. She married again, and infected her second husband. After her death the second husband marries again, and communicates the disease to his second wife. See also cases by Dr. Villemin *L' Union*, 1868.

† See Chauveau's experiments by feeding calves with tubercular matter. "Gazette de Paris," p. 47, 1868.

‡ "Baxine Tuberculosis in Man." London, 1881.

the chief points that have been advanced to support the communicability of tubercular affections is given by Dr. Klein in the "Practitioner" for August 1881. He refers to a large number of illustrative cases. It must be confessed, however, that the "virus from without" theory places its upholders in certain awkward positions when they proceed to discuss the clinical bearings of the case in greater detail. Cohnheim,* for example, maintains that tuberculosis of the air passages is due to a tuberculosis virus that has been inspired, and that has been inspired, and that of the intestinal canal to like matter swallowed. So far one can follow him: but when he attempts to explain primary tuberculosis elsewhere it must be confessed that his explanations are at the least fanciful. Thus he conceives it possible that in meningeal tuberculosis the virus may enter the skull from the nose *viâ* the cribriform foramina; and in cases of primary tubercular disease of the kidney it is needful to assume the existence of a virus in the blood, which virus is in time excreted by the kidney, and so becomes localized in the gland. In the same way, in tuberculosis of bone, where a primary injury is so common, he supposes that that injury attracts a virus already in the blood by the inflammation it excites. It must be allowed that this is all but the wildest conjecture.

Some other recent experiments as to the inoculability of tubercle bring out fresh aspects of this vexed question. These experiments I will briefly allude to. Their real significance can hardly be yet discussed, as the whole matter is still *sub judice*. Klebs, Hueter, Schüller, and others, have endeavored to show that the tubercle virus is a micro-organism, and that can be best developed for inoculation purposes by what is known as "fractional culture." A piece of tubercular tissue is treated in a certain manner, and in the fluid about it a number of these micro-organisms (*spaltpilze*) develop. If the fluid that contains these organisms be injected into the body of certain animals both local and general

---

* "Die Tuberculose vom Standpunkt der Infectionslehre." Leipsig, 1880.

tuberculosis is produced, although the injected matter contained actually no trace of the original piece of tissue employed.* Deutschmann † has, however, more recently repeated these experiments, employing a somewhat different mode of procedure, and he asserts that with a fluid containing micrococci identical with those described by Klebs and others, he obtained only negative results, and in no case succeeded in inducing tuberculosis. The whole question therefore must be considered as still in a very unsettled condition.

Schüller's experiments‡ show also that certain tubercular products are much more active as inoculating agents after cultivation of the specific micrococcus. Thus lupus tissue, when subjected to cultivation, furnished a fluid that on inoculation produced both local and general tuberculosis; whereas inoculation with fresh lupus matter either led to no results, or, as in M. Kiener's cases, to but a slight local lesion. Then, again, in another series of experiments, Schüller introduced tubercular matter into the trachea of an animal without inflicting any wound. He then contused one of the creature's joints, and the result was a white swelling or tubercular joint disease. A similar result also followed joint contusion in some cases where non-inoculated animals had been simply living in contact with animals that had been rendered tuberculous. These apparently healthy animals developed a tubercular joint affection. A like inquiry, however, in animals that had not been so associated, led to none but a passing local disturbance. Dr. Baumgarten§ recently introduced into the anterior chamber of the eye of a healthy rabbit a drop of fresh blood taken from a rabbit afflicted with general artificial tuberculosis. In two or three weeks an eruption of tubercles appeared on the iris just as in Cohnheim's cases.

* An epitome of these experiments will be found in Dr. C. Hueter's "Grundriss der Chirurgie." Leipsig, 1880, vol. i. p. 270.
† "Centralblatt f. Med. Wissensch.," No. 18, 1881. p. 322.
‡ "Experimentelle und histologische Untersuchungen, uber die Enstehung und Ursachen der Scroph. und Tuberk. Gelenkleiden.' Stuttgart, 1880.
§ *Centralblatt f. Med. Wissensch*, No. 15, 1881, p. 274.

CHAPTER V.

A DEFINITION OF SCROFULA.

I would define scrofula as a tendency in the individual to inflammations of a peculiar type, the distinctive features of such inflammations being as follows:—They are usually chronic, apt to be induced by very slight irritation, and to persist after the irritation that induced them has disappeared. The exudations in these processes are remarkable for their cellular character and for the large size of some of those elements. Such exudations also show a remarkable tendency to resist absorption and to linger in the tissues, the affected area becoming rapidly non-vascular. Among the common products of these inflammations are giant cells, and, if a certain stage of the process be reached, tubercles. The tendency of the process is to degenerate, not to organize, and the degeneration usually takes the form of caseation. At the same time these inflammations have a tendency to extend locally and infect adjacent parts, and their products present certain peculiar properties when inoculated upon animals. Lastly, a great feature of all these processes in this — they tend to commence in and to most persistently involve lymyhatic tissue: an implication of this tissue being a conspicuous feature in every case of scrofulous disease.

The tendency to this peculiar form of inflammation may be called, if so wished, a diathesis, or, more definitely, the scrofulous diathesis.

I do not propose to discuss here all the points of this definition. Some of the special features of scrofulous inflammation have often been considered in dealing with the subject of tubercle, and among these the inoculability of the products of such inflammations has been referred to. The histology of the process, the peculiarities of its exudations, its mode of decay, and its relation to adjacent tissues, will all be fully treated of in the section that deals with the pathology of lymphatic gland disease. There is only occasion, therefore, to consider in the present place these few remaining features, viz., 1. The chronicity of the scrofulous process. 2. The slight irritation that may induce it. 3. Its tendency to extend locally and by continuity of tissue: and, 4. The remarkable and constant manner in which it involves lymphatic tissue.

1. The chronicity of all scrofulous manifestations is well known. The gland affections are slow in their progress, often extremely slow, so that their duration may be estimated more often by years than by months. In like manner, the classical bone and joint affections of the scrofulous are essentially chronic. The same leisurely course can be observed in the skin eruptions and in the disorders of the mucous membranes. If acute inflammatory changes do occur in the subjects of scrofula, such changes are nearly always accidental, and may be regarded as complications of the process that are by no means either usual or necessary. Indeed, those who present scrofula in a marked degree seem singularly little prone to acute inflammations of any kind. Their tissues seem to react rather with the utmost torpidity under the inflammatory process. And in other affections that are only incidentally associated with inflammation a like tendency to a chronic action can often be observed. For example, I have in one or two cases seen herpes zoster in a scrofulous child assume a very tedious course, and induce an amount of sluggish suppuration not usually met with in that neurosis.

2. In treating of the gland affections in scrofula I shall draw attention to the trifling character of the peripheral lesion that is often sufficient to set up the disease in the absorbents. A slight ulcer within the mouth, defects in dentition, a trifling eczema behind the ears, an ophthalmia, are all sufficient to induce a considerable gland disorder; and that disorder will persist, and indeed progress, after the initial disturbance has entirely disappeared. So in bone and joint affections and in spinal caries one is often surprised at the slight traumatism that may induce a very severe and extensive morbid change. A trifling exposure to cold, that in a healthy child would have little or no effect, suffices often to arouse an obsti-

nate conjunctivitis, or catarrh, or ulceration of a mucous membrane, that persists, aud for a long while resists treatment. Cold abcess, again, often makes its appearance after the most insignificant injuries; and, indeed, a large number of scrofulous maladies have an origin so obscure that they are regarded as spontaneous in their nature. In the subjects of scrofula there appears to be, indeed, a remarkable vulnerability of tissue, a strange proneness to lapse into a condition of disease after irritations usually regarded pathologically as of no moment. This vulnerability of the tissues has assumed a conspicuous place in the descriptions of scrofula advanced by certain pathologists. Virchow assigns to it a prominent position in the pathogenesis he proposes for the disease, and many others have regarded it as a still more essential factor. Recently M. Paul* has made this inherent weakness of scrofulous structures a feature in the diagnosis of the affection, and would have us recognise a class of scrofulous person from the manner in which the puncture in the ears for ear-rings takes on unhealthy action, and leads to linear scars, fissures, and similar deformities. There are certain animals, such as the rabbit and guinea-pig, in which caseous (or, as it may be fairly called, scrofulous) inflammation is very common; and it is remarkable that in these animals that form of inflammation is often induced by injuries of a comparatively trifling nature. Too extensive inferences must not be drawn from the vulnerability of tissue in the scrofulous. It is not, for example, to be inferred that any constant, or even common relation exists between the severity of the initial lesion and that of the subsequent disease. It is only maintained that the tissues of the scrofulous are apt to react with an almost characteristic readiness to disturbing causes that in the healthy would rank as insignificant.

3. The tendency of the scrofulous process to extend locally is a very distinct feature of the disease, and is apparent in most of its manifestations. In lupus this

---

*Sur un nouveau Signe de la Scrofule fourni par les Boucles d'Oreille. "L'union Medicale," Feb. 26, 1881, et seq.

tendency is very conspicuous, and forms, indeed, one of the characteristics of that affection. The morbid process can be observed to deliberately extend, and to invade the adjacent parts in a progressive manner. In some cases this extension may be considerable and very widespread. I have now under my care a girl of 16 with lupus non exedens, that has involved the whole of the right upper extremity, and has produced such contraction as to render the limb useless. The process has spread into the neck, and involved nearly the whole of both sides of it, and over one of the lower extremities a like extensive lupoid change has occurred in the integuments. One is familiar also with the very deliberate manner in which certain scrofulous ulcers will extend in spite of any but the most vigorous treatment, spreading not infrequently in a manner as distinct as is observed in certain ulcers in the tertiary syphilitic. In cases also where the skin becomes undermined about a sinus—such a sinus as may form after the breaking of a gland abscess—it is remarkable to observe how in some instances that undermining will extend. The undermined integument is thin and purplish, and where it joins the healthy skin a little subcutaneous induration can often be felt. This induration represents the spreading scrofulous process, and it is by its ultimate breaking down that the process of undermining gradually extends. Retained pus may assist in bringing about this condition, but if so it acts only as a very feeble auxiliary; for the undermining of the skin may extend in cases where the pus has perfectly free vent, and where elaborate drainage is carried out. Rather is it due in the majority of cases to a gradual extention of a scrofulous process in the subcutaneous tissue, fostered and augmented, no doubt, by unhealthy changes already in action; but yet to such an extent is it a new morbid process that the treatment of skin so undermined becomes very different from the treatment of that sapped by simple suppuration. Lannelongue has shown in a very exhautive manner how a cold abscess extends, how its wall presents an active scrofulous process and an abundant deposit of perfect tubercles, how this wall gradually degenerates and breaks down into the abscess cavity on the one hand, while, on the other, it slowly invades the adjacent tissues without line or barrier, and how it is by the extension of this invading wall that the abscess cavity enlarges.

The same condition holds good with regard to the bone affections of the scrofulous, and is well seen in caries. Here the diseased process spreads leisurely and deliberately, unlimited by any barrier of healthier action. The morbid condition shades off so gradually into the as yet unaffected bone, that it is often difficult to say where disease ends and sound bone begins. M. Lannelongue has compared the spreading of this disease in bone to its spreading in the case of cold abscess, and regards the two examples of tissue invasion as identical. In speaking of gland affections I shall have occasion to deal at some length with this feature of local extension in scrofulous disease, and shall show how the process may creep from one gland to another independent of any new or abiding source of peripheral irritation; how one gland may, as it were, infect another, and how the path of infection is along connecting lymphatic vessels.

Cases are sometimes met with where local extension by means of the lymphatics has been very considerable, and where disease in widely separate parts has been connected by continuity of tissue. For example, Dr. Pye Smith,[*] under the title of " Primary Caseous Degeneration of Lymphatic Glands," describes the case of a woman, aged 47, in whom caseous inflammation commenced in the lymphatic glands of the mediastina, set up apparently by a bronchitis. The process spreading upwards from the bronchial glands reached the neck and involved both sides of it. Extending in the opposite direction it reached the glands of the omentum and mesentery, so that the whole of the diseased parts formed one continuous series. Dr. Hilton Fagge[†] records a like case occurring in a woman aged 35. Here

---

[*] " Trans. Path. Soc.," London, vol. xxvi. 1875, p. 202.
[†] " Trans. Path. Soc.," London, vol. xxv. 1874, p. 235; see also similar case under title of " Fibroid Disease of Heart," on p. 72 of same volume.

the gland mischief commenced in the left groin, and from that spot a cotinuous chain of large caseous glands were traced up along the front of the spine into both sides of the neck. The spreading here by continuity of tissue was deliberate and definite. Some of the gland masses, especially those about the trachea, were moreover very large. Dr. Goodhart* records a case in some respects still more interesting. The patient, a man aged 22, suffered from pulpy degeneration of the right knee, that ultimately required amputation. Death resulted, and at the post-mortem it was found that the glands in the right Scarpa's triangle were involved, and from thence a continuous chain of diseased glands could be followed up along the psoas muscle, along the front of the spine, about the root of the lungs, and so up into the neck, especially implicating the glands of the right side. Here, then, there existed a continuity of disease from the knee to the posterior triangle of the neck. Little more need be said upon this subject. Suffice it to state that the same tendency to local spreading or infection is exhibited in other scrofulous affections, notably in the scrofulous ulcer of mucous membranes and in that local tuberculosis, that quasi-scrofula that precedes the eruption of general tuberculosis in the inoculation experiments in animals.† On this matter, considered in its entirety, we would fully endorse the observation of M. Kiener as to scrofulous affections, that "chaque foyer exerce sur les tissus environnants une action infectieuse de voisinage, d'où résulte la formation de nouveaux foyers." ‡

4. The marked implication of lymphatic tissues in all the manifestations of scrofula is the feature which, of all isolated attributes, I would urge to be the one most significant of the process From the earliest days of medicine, scrofula has been associated in some way

---

* "Guy's Hospital Reports," vol. xviii, 1873, p. 401.
† See "The Artificial Production of Tubercle," by Dr. Wilson Fox. London, 1868.
‡ *L' Union Medicale*, Feb. 22, 1881, p. 320.

or another with the lymphatic system; and such association is no matter of wonder when the remarkable tendency to lymphatic gland enlargement in the scrofulous is borne in mind. The older authors conceived some humor or acrid matter in the lymph that caused it to coagulate in the glands. Others maintained that fluid to be of too great consistence. Others, that the lymphatic vessels were at fault, or the glands were so ill constructed as to prevent the lymph from passing through. In more recent times, the association of scrofula with the lymphatic tissues has assumed a less indefinite outline. Bell and Hufeland considered that scrofula was due essentially to a certain weakness or atony of the lymph system. Virchow adopts a very similar view, and ascribes the disease to a great extent to an incompleteness in the structure of the lymphatic glandular apparatus. Birch-Hirschfeld,[*] in alluding to this latter tissue weakness, considers that it may not be unlike that hereditary defect in the vascular apparatus that marks hæmophilia. Villemin [†] regards scrofula as due to a morbid irritability of what he terms the lymphatico-connective system—a tissue system composed of the connecting structures of the body, and the system of lymphatic channels with which those tissues are in such immediate and direct relation. He ascribes the chronicity of the superficial manifestations of scrofula, their tendency to extend, their inclination to involve deeper parts, to implication of this system of tissues, and regards such implication as the essential feature of the whole process. Like views have been advanced by others. To come, however, to matters more of detail. The first point to be noted is the great tendency to gland enlargement in all strumous disorders. This fact alone establishes a remarkable alliance between scrofula and the anatomical element, lymph tissue. But if other scrofulous affections are observed in detail, the same alliance, and indeed a still closer one, becomes obvious.

---

[*] Ziemssen's "Cyclopædia of Medicine," vol. xvi, p. 763.
[†] "Scrofulisme et Tuberculose." *L'Union Medicale,* March 29, 1881.

One of the commonest manifestations of scrofula is the enlarged tonsil, and it is needless to observe that the tonsil is simply a mass of lymphoid or adenoid tissue. Scrofulous pharyngitis is merely a caseous inflammation of the lymphoid tissue of the pharynx,* the scrofulous ulcer of the intestine has its original seat in the adenoid structures of the gut, the tubercular ulcer of the larynx begins in the lymph follicles of the part,† and a like intimate structural relation has been shown by Rindfleisch ‡ to exist in tubercular ulcers of the bronchial mucous membrane. And in the other affections of mucous surfaces common in the scrofulous, such as ozœna, coryza, and vaginitis, there are strong reasons for believing that a considerable implication of the lymphatic structures exists. Mr. Greig Smith, § in an interesting and most valuable contribution to the pathology of strumous bone disease, alludes to the important part the red marrow plays in these affections; and holding in mind the intimate connection that undoubtedly exists between this marrow and the general lymphatic apparatus of the body, he ventures to speak of such bone disease as essentially a lymphadenitis. When one comes, however, to microscopic investigation, the evidence as to this relationship proves more than sufficient. I would urge that lymphatic structures of some kind, no matter whether vessels, channels, or adenoid tissues, are essential to the formation of tubercle, and that this tissue forms as much the basis of tubercle as epithelium does of epithelioma. Lymphatic tissue is, of course, almost universal in its distribution, but it is the very prominent implication of that structure that is so marked a feature in tuberculosis. If the tubercle formed in the artificial tuberculosis of animals is to be regarded as in any way a typical product, then must it be acknowledged that its origin is almost exclusively from lymph tissue.

---

* Wendt. "Ziemssen's Cyclopædia," vol. vii, p. 75.
† Dr Curnow. "Gulstonian Lectures." *Lancet*, vol. i, 1879, p. 510.
‡ *Loc. cit.*, "Ziemssen's Cyclopædia," p. 663.
§ "Reprint from the British Royal Infirmary Reports," 1878-79, p. 99.

Such tubercles, when met with in the lungs, commence merely as nodular enlargements of the adenoid tissue, that normally exist around the bronchi and the blood-vessels of the part. The whole process concerns the lymphatics. If the matter be introduced into the pleural cavity, the first evidence of the pleuritis induced shows itself about the surface lymphatics. If the so-called tubercular process extends to the lungs, that extension is by the lymphatics (Klein). If it extends within the lungs, it is by the lymphatics. Wilson Fox has shown the same extraordinary implication of lymphatic structures, and indeed in his experiments with injections into the subcutaneous tissues, the whole progress of the malady is to be followed by following the lymphatics. He strongly insists upon the lymphatic structure of tubercle, points out its origin from lymphatic tissues, and suggests that it may be due to some peculiar morbid disposition in those tissues.*

The origin of tubercle from the perivascular lymphatics has, I think, been very clearly demonstrated, especially with regard to the blood-vessels of the pia mater. Cornil and Ranvier † give an excellent drawing to show the development of tubercle from a lymphatic vessel in a case of tubercular ulcer of the intestine. Rindfleisch ‡ has shown how in the lung the tubercular ulcer of the bronchus spreads locally by invading the lymph channels of the part. In dealing subsequently with the giant cells of tubercle, I shall endeavor to show that these bodies can only appear when lymphatic channels or tissues are provided as an anatomical basis, and shall indeed hope to prove that they are merely peculiar lymph coagula.

There are other points eminently suggestive of a serious implication of the lymphatic apparatus in the scrofulous. Dr. Grancher § was fortunate enough to

---

* *Loc. cit.*, p. 29.
† *Loc. cit.*, p. 634, fig. 255.
‡ *Loc. cit.*, p. 663.
§ "*Loc. cit.*, Dictionnaire Encyclopedique," p. 311.

obtain sections of the hypertrophied upper lip from a scrofulous child. This deformity, which is considered by some as very typical of scrofula, is no doubt due merely to irritation of the part by previous local mischief, usually by unwholesome discharges from the nose. This hypertrophy showed on examination merely a great dilatation of the lymphatic capillaries of the subdermic tissues, with thickening of their walls. In some places the greatly distended spaces were partly blocked by an accumulation of lymph and coagulated fibrin. Dr. Curnow, * speaking of inflammatory affections of the lymphatic vessels, says, " With reticular lymphangitis I am inclined to include the acute swellings of the lips and tip of the nose, which is so common in strumous people, and the red and painful patches in the vicinity of eczematous eruptions on the nose, lips, and ears, inasmuch as it is only where lymphatic networks are especially abundant that such œdematous swellings occur." Hueter † considers that the sodden and pasty condition of the skin seen sometimes in strumous subjects may be due to a permanently dilated state of the lymphatic vessels, and such a suggestion would appear extremely probable. Occasionally, in delicate scrofulous children there is a strange tendency for wheals to develop on the most trivial provocation. Dr. Thomas Barlow‡ records the case of an infant in this condition in whom a slight scratch or even friction of the skin brought out a wheal almost immediately. That these wheals are to be explained by an injury to greatly enfeebled lymphatic vessels and channels appears to be very probable, and to be aptly compared to the hæmorrhages that may occur from slight lesions in certain enfeebled conditions where a weakness of the vascular capillaries is imagined to exist.

What is actually the anatomy and physiology of the lymph apparatus in the scrofulous is still a matter of

---

* " Loc. cit.," p. 508. Dr. Curnow describes the lymphatics as being especially numerous at parts where skin join. mucous membrane.
† " Grundriss der Chirurgi," 1881, p. 266.
‡ "Trans. Clinical Soc.," London, vol. x. 1877. p. 197.

conjecture, but that this apparatus presents a strange vulnerability, a remarkable tendency to encourage and invite disease in those who present the scrofulous diathesis, must, I think, be allowed.

CHAPTER VI.

SCROFULA AND PHTHISIS, AND THE ANTAGONISM BETWEEN SCROFULOUS DISEASES.

The relation between these two affections has been a subject for endless dispute, and is still under discussion. The matter has been debated from both a clinical and a pathological standpoint, and it must be confessed that the most opposite opinions have been supported by no small amount of valuable evidence.

I might at once state the particular opinion I venture to urge on this matter. I believe that scrofula and phthisis are due to the same morbid process, that phthisis may be regarded as scrofula of the lung in like manner as a scrofulous lymphatic mass may be regarded as phthisis of a gland. I would acknowledge no relationship between the two other than that of their identity and the actual sameness of the morbid action in the two diseases; and would entertain no such alliance between them as that of cause and effect, or soil and seed, of primary disease and secondary disease, all of which relations have from time to time been insisted on. The observation of a few clinical facts are sufficient to impress one with the intimacy that exists between these two diseases. They both occur not infrequently in the same kind of delicate person. I do not mean that such individuals have a distinct physiognomy. They are classed with the vague mass of the delicate, but present certain vague marks of frail health that are common both to those who exhibit scrofula and to those who are phthisical. The description of a child with "a phthisical tendency," as given in many text-books, very fairly accords with what is known as the sanguine or erethic form of scrofula. Then, again, in the etiology of the two affections there is a remarkable unanimity when tendencies that may be regarded as purely local are excluded. The same general causes that predispose to phthisis, predispose to scrofula, a fact that Ruehle* has pointed out in some detail. The same observer also states that, as a general rule, where scrofula is very prevalent phthisis is also common. To this latter rule there are of course exceptions; but I think that the different proportions in which the two diseases exist in certain parts of the world can be explained by conditions that in one case lead rather to the surface ailments of scrofula, and in the other to general pulmonary disorders.†

A simple example of a common cause, leading in one case to scrofula and in another to phthisis is afforded by measles. It is well known that measles is frequently followed by cervical gland enlargements, induced, it is supposed, by the coryza that accompanies it, and that it may act as the exciting cause of more extensive strumous disease in those already predisposed to scrofula. Measles also—as Ruehle ‡ has shown—has often an intimate concern in the etiology of consumption, and may be considered in this instance to act through the bronchitis with which it is attended. Here, then, the same malady in two different situations excites the same disease in corresponding parts; for it is hard to conceive special conditions for the mucous membrane of the lung that do not also hold good for the mucous lining of the nose and pharynx. Moreover, in the matter of heredity these two disorders are often seen to be

---

* "Ziemssen's Cyclopædia of Medicine, Art. Pulmonary Consumption," vol. v. p. 496.

† Dr. Thaon, who maintains the identity of scrofula and phthisis, and speaks of phthisis and scrofula of the lung, has drawn attention to certain local and climacteric influences that lead to the unequal distribution of scrofula and phthisis in certain parts. *L'Union Médicale*, vol. xxxi. 1881. p. 40.

‡ "Loc. cit.," p. 504.

interchangeable. A phthisical parent may beget scrofulous children, and a scrofulous parent phthisical offspring. Or in a given family with a history of what might be termed a tubercular taint, some of the children may become scrofulous, while others are phthisical, and the rest perhaps present simply that delicacy of health that we know might lead to one or other of those diseases.*

When, in the next place, one comes to compare the morbid changes in consumption with those in scrofula, to compare a phthisis of the lungs, on the one hand, with (let us say) a scrofulous gland on the other, I think it must be owned that the resemblance is very close. I ventured to define scrofula as an inflammation presenting certain distinctive features, and I would maintain that these very features form also the distinctive attributes of the phthisical process. If the definition already given to scrofula can apply also to phthisis, then phthisis may be thus described. It is a process usually chronic, apt to be induced by very slight irritation, and to persist after that irritation has disappeared. Its exudations are remarkably cellular in character, prone to resist absorption, and to linger in the tissues. Among the common products of the process are giant cells, and in certain cases tubercle; the affected districts in any case becoming rapidly non-vascular. The tendency of the process is to degenerate, and such degeneration usually assumes the form of caseation. The process, moreover, has a tendency to extend locally and infect adjacent parts; its products also produce certain results when inoculated in animals.

This is, I believe, no imperfect description of phthisical change, and yet it is merely a repetition of the terms used in defining scrofula. The naked eye changes in the parts are very similar: The lungs and the gland become affected in certain spots, which spread and fuse. Caseation occurs. The tissue breaks down, and a cavity is produced. The destructive action is,

---

* Tyler Smith well describes such a family in his work—"Scrofula: its Nature, Causes, and Treatment." London, 1844, p. 6.

for a time at least, not limited by a barrier of healthier action. In both the gland and the lung, however, the caseous mass may be encapsuled or become cretaceous, or the whole process may be associated with a great development of fibrous tissue (fibroid phthisis), or lastly, cure may follow by any one of these less frequent changes. The cavities in a phthisical lung appear to me in no way unlike the broken down spaces in articular osteitis, or the excavations in the testicle in scrofulous orchitis, or the purulent cavity beneath the skin due to the breaking down of what is known as a scrofulous gumma. The implication of lymphatic tissue in phthisis will be immediately alluded to, but here I can but draw attention to the frequent implication of the bronchial glands in phthisis, and the absolute identity of such glands with those more superficial organs usually denoted scrofulous. Is there any lung affection (apart, perhaps, from certain rare new growths) where the glands are so constantly involved as they are in phthisis? I imagine not; and this fact alone appears to be most significant. Moreover, the clavicular and axillary glands may be involved in cases of phthisis, and present on removal precisely the characters of scrofulous glands.

In the third place, brief notice may be made of the resemblance that exists in the histology of the two affections. Phthisis, it is very generally allowed, commences by an inflammatory process, usually a catarrhal pneumonia. Whether the catarrh is simple, as Niemeyer would urge, or from the very first specific, as Ruehle and others insist, is a matter of little moment, and a question hardly to be settled by reference to the histology of the disorder. This catarrh is, as Niemeyer* observes, equivalent to the initial inflammation of the scrofulous, to the inflamed pharynx or conjunctiva that leads to gland disease, or to the catarrh that deepens into a scrofulous ulcer. Its exudations are not removed, and soon changes occur in the walls of those alveoli whose cavities are already filled with catarrhal products.

---

* " Text-Book of Practical Medicine," vol. i,, 1873, p. 212.

I would point out that these changes considerably involve the lymphatics. The exudation in the alveolar wall consists at first mainly of lymphoid cells. These, as Grancher * says, are at first ranged in linear rows, and occupy in fact the interfascicular spaces of the tissue; such spaces, it is known, are merely lymph channels. Then, again, Rindfleisch † connects these changes in the alveolar wall with certain changes in the adenoid structures about the smaller bronchi and arteries, and one knows that the perivascular adenoid tissue is at least in direct continuity with the lymphatic channels of the alveolar wall (Klein). If the phthisical process is rapid—as in Phthisis florida—degeneration may ensue before any giant cells or tubercles have been met with (Ruehle), the exudations and the altered lung tissue simply becoming caseous and rapidly disorganized. A precisely like condition occurs in certain glands that caseate and break down with great rapidity. In other cases of phthisis giant cells or tubercles may appear, and their advent be followed by caseation and the usual mode of ending. Like conditions are common in the glands. Lastly, fibroid thickening is often conspicuous in the lung, especially in cases of slow progress and of little intensity; and such a change, I shall show, is also met with in some glands, although, owing probably to the better blood supply of the lung, it is more common in that organ than in the lymphatic glands.

Lastly, scrofula and phthisis may occur in the same person, and *à propos* of this one meets with some singular conflicts of opinion. It will be acknowledged, I suppose, that scrofulous patients may become phthisical, and that phthisical patients may present scrofula, but can we go so far as to say with Lugol ‡ that "the natural death of the scrofulous is by consumption," or with Hamilton § that "at least 9 in 10 of those who die

---

* "Tuberculose Pulmonaire, Archives de Phys.," 1878.
† "Ziemssen, loc. cit.," p. 652.
‡ "Researches and Observations on the Causes of Scrofulous Diseases. Translation. London, 1844, p. 48.
§ "Observations on Scrofulous Affections," 1791, p. 27.

of consumption are scrofulous subjects," or even with Ruehle * that scrofula is one of the chief predisposing causes of phthisis," or with Rindfleisch † that "tuberculosis (including phthisis) hardly ever occurs except in scrofulous persons?" I imagine not. There appears to be an impression that if scrofula and phthisis are in any way allied phthisis should be a common cause of death in the scrofulous. I know of no two allied diseases where the impression is so vividly maintained as it is in the present instances, and I am unable to see why such a mode of death should be considered as necessary to support any such alliance. Yet it is one of the most prominent arguments of those who advance the identity of the two disorders. I would, however, say that while scrofula and phthisis are manifestations of the same morbid process, I am nevertheless convinced that phthisis is by no means even common in the scrofulous, and that the bulk of such patients do not die of pulmonary consumption. If scrofulous patients are so prone to die of phthisis as some maintain, then phthisis must be a terribly prevalent disease, for there are few general maladies more widespread than is scrofula; and as it is not maintained that any relation exists between the severity of the strumous lesion and the proneness to consumption, all cases may become phthisical. Moreover, apart from this, if such a relation did exist one would certainly expect to find evidences of scrofula in a large percentage of phthisical patients. But investigations lead to quite an opposite conclusion. Through the kindness of Dr. F. J. Hicks, medical officer to the Brompton Hospital, I was enabled to carefully examine 57 cases of phthisis for evidence of any present or past manifestations of scrofula. Out of the 57 cases I found traces of scrofula in only 7 instances, and some of these included trifling phases only of the disease.‡

---

* "Loc. cit.," p. 604.
† "Loc. cit,," p. 639.
‡ The 57 cases included patients of very different ages, and were taken simply in the order in which they came in the wards—39 were females and 18 males. Of the 7 cases that showed any evidence of

In 332 cases of phthisis examined post-mortem by Mr. Phillips he found " scrofulous scars " in 7 only.* This meagre result is of course explained by the fact that Mr. Phillips looked merely for one particular outcome of scrofula, viz., scars due to glandular abscess. In addition to his own observations he has collected like statistics by other surgeons. These include 1,078 phthisical patients who were examined post-mortem. In only 84 instances out of this large number was there any evidence of cervical gland disease. Mr. Kiener† relates the after history of 87 cases of scrofulous bone and joint disease with the result that out of this number 6 only died of tuberculosis of the lungs or meninges. According to Villemin‡ " a considerable number of the tuberculous (including the phthisical) show neither vestige nor souvenir of scrofula," and further, he states his opinion that persons attacked by scrofula in their childhood do not more often become tubercular than do other persons. Grancher § also strongly opposes the attempt to make out that all scrofulous affections tend to end in phthisis, and will not allow that it is the ultimate manifestation of those diseases.

I maintain, then, that scrofula and phthisis are identical in their nature, but that phthisis is by no means a common complication—either immediate or remote—of the former disease. The explanation I would offer of this supposed discrepancy, and indeed of the general

---

scrofula, 5 were females and 2 males; 3 patients besides had enlarged tonsils. In several a slight enlargement of one or two glands in the neck was detected, but in each instance it was associated with some recent trouble in the pharynx or larynx dependent on the phthisis. The 7 cases with scrofula were as follows; F., aged 39, caries of carpus, nine months. F., 24, gland abscess in neck, seven years ago. F., 24, ozœna three years ago, cervical gland disease two years ago (no trace now left of either affection). F., 35, caries of sternum two years ago; sinus existing. F., 19, extensive gland disease in neck, of six months' standing. M., 13, foot removed ten years ago for disease of ankle joint. M., 25, scrofulous epididymitis.

* " Scrofula and its Treatment." London, 1846, p. 75.
† " L'Union Medicale, loc. cit.," p. 319.
‡ " Ibid.," p. 497.
§ " Loc. cit., Dict. Encyclop.," p. 325.

relations between struma and pulmonary consumption, is the following.

There is a decided antagonism between scrofulous diseases of all kind. If a patient has one severe or even well-marked manifestation of scrofula he is not likely to develop another strumous disease at the same time. Indeed, the particular scrofulous malady any given patient presents would appear to protect him from any other outcome of the disease for at least the time being. The records of the Margate Infirmary for scrofula illustrate this statement in a very distinct manner. From fully recorded details of 509 cases of scrofula in this Institution I obtain these results*—248 of the cases were males, and out of this number only 27 patients presented more than one grave manifestation of scrofula at one and the same time. In the remaining 221 cases there was only one grave outcome of struma in each patient. 261 of the cases were females, and among this number there were but 29 affected with more than one scrofulous malady at the time they were under treatment. Placing the male and female cases together, it will be seen that out of the 509 instances of scrofula only 56 patients presented concurrent scrofulous affections. I must explain that these 56 cases include only such as present distinct and separate scrofulous diseases. I have omitted cases where two affections in the same patient stood to one another in the relation of cause and effect. In such cases one of the affections was *always* glandular, and as examples I must cite: ophthalmia with enlarged neck glands, ozœna with a like complication, ulcers of the foot with inguinal bubo, and similar instances. All such cases I have excluded, except one or two where the gland disease had assumed considerable proportions. I have also excluded cases where a patient presented evidences of a cured disease and an active disease at the same time. The 56 cases include

---

*These cases were taken without selection, from records kept by Mr. C. B. Waller, now of Sydenham, and Mr. W. K. Treves, of Margate, during the periods they held the office of resident surgeon to the Infirmary. These records are singularly complete, and form an invaluable series of cases.

instances of disease such as these—caries of lumbar spine, with gland disease in the neck, disease of the femur and humerus in the same patient, disease of two or more joints at the same time, or association of such bone and joint affections, with glandular enlargements in distant parts.

I think these results are sufficient to show that there is an antagonism between scrofulous affections, and that it is not usual for two grave manifestations of the disease to be active at one and the same time. Moreover, it is common to observe one strumous disease subside or improve when another becomes manifest. Cases like the following illustrate this fact :—

A little girl, aged 14, with a history of phthisis in her family, had presented the following succession of diseases. At the age of 6 she suffered from caries of the tarsus, leading to amputation of the foot in a year's time. When 8 years old she became afflicted with multiple subcutaneous abscesses. These in time healed, and at the age of 12 she began to present cervical gland enlargements, which had commenced to suppurate when she came under notice. Female, aged 23. When 13 years of age she was the subject of considerable gland disease in the neck. These enlargements persisted until the age of 19, when she was attacked with a lupus of the face. On the appearance of the lupus the gland affection began at once to subside. The following case, reported by Birch-Hirschfeld, may well be quoted here: —" I have had for the last three years under my treatment a scrofulous girl, 12 years old, who from time to time is troubled with ophthalmia, coryza, and eczema of the face of great severity. In this case I could repeatedly observe painful tumefaction to a considerable extent of the cervical lymphatic glands (especially near the angles of the jaw), whenever these phenomena receded. As soon as the first symptoms exacerbated, an evident remission of the glandular swelling, and especially of the pain in it, took place, and, strange to say, the general condition always improved with this remission. That this was not a mere coincidence is proved by the repeated occurrence of this alteration of symptons, it being observed not less than five times in the course of a single year." * I have notes also of cases where a marked improvement occurred in cervical gland disease on the patient becoming the subject of hip mischief or an extensive bone affection, a change hardly to be expected when one remember the further deterioration of health such joint and bone affections must imply.

Now classing phthisis with scrofulous diseases I would maintain that the same antagonism is observed in its life history as obtains in diseases commonly designated scrofulous. An individual with scrofula is no more likely to become phthiscal than a patient with one grave scrofulous disorder is likely to develop another. Cases of course do occur, as already allowed, where scrofula and consumption are met with in the same patient, but such instances are not common. Many of the old authors recognized this antagonism between what they termed external and internal tuberculosis, and asserted that " so long as gland disease is active the lungs may be regarded as safe, whereas if phthisis supervenes the glands decrease rapidly, and may even disappear." † Recent writers have approached the subject with somewhat more caution, but there are not a few who allow this reciprocity in the two affections. Mr. Holmes, for example, remarks, " moderate enlargement of glands in patients with a family history of phthisis is often considered as a derivative or preservative against visceral mischief, and I must say that I incline to this opinion." ‡ Dr. Walshe, speaking of the relation between scrofula and phthisis, observes, " The external lymphatic system on the whole rarely undergoes tuberculization in the phthisical state. An antagonism, not absolute but tolerably well marked, seems to exist between the external and internal tuberculizing processes. In corroboration of this I have

---

* " Loc. cit.," p. 781.
† " On Scrofulous Disease of External Lymphatic Glands," by T. Balman. London, 1852, p. 417.
‡ " Surgical Treatment of Diseases of Infancy and Childhood." London, 1868, p. 637.

known the cervical and axiliary glands greatly enlarged in phthisical people, rapidly fall to the natural size without suppuration or symptom of any kind, while pulmonary tuberculization rapidly advanced."* If phthisis is invited as it were by any frailty of health, and if it is so prone to attack the scrofulous as some would have us believe, why is the disease not more frequent in the subjects of hip diseases, who lie bedridden for months in the feeblest state of health, and die at last perhaps of amyloid degeneration of their viscera? Why is it not more common in the subjects of angular curvature, with the associated deformity of their chests? and why do we not more often meet with it in cases of cervical gland disease, where the scrofulous masses extend down the trachea, on the one hand, or actually reach to the pleura on the other? Such patients are, I believe, protected by the antagonism that exists between scrofulous diseases. Many facts in the histories of phthisical and scrofulous families illustrate this point. As examples, these cases may be cited as a few of many:—

A female, aged 27, presented herself in my out-patient rooms with extensive strumous enlargements in the neck, that had troubled her more or less from childhood. She presented no traces of phthisis. Her father has been phthisical. She was the youngest child but one out of a family of seven. Of these the two elder died marasmic in infancy; the third, a male, died at 21 of phthisis; the fourth, a male, died at 22 of phthisis; the fifth, a female, died at 30 of phthisis; and the youngest member of the family, a male, aged 20, is at present in good health. The patient is the only one of the family who presented symptoms of scrofula, and it would appear that this disease had saved her so far from the fate of her brothers and sister. A case as marked was that of a woman, aged 46, also an out-patient of mine, with suppurating glands in the neck. This gland disease had troubled her since she was 11 years old, and suppuration had occurred so many times

---

* "Diseases of the Lungs." London, 3d ed., 1860.

that her neck was covered with scars of various dates. Her father was a healthy man, and had been killed by an accident. Her mother died of phthisis. She had 5 brothers and sisters, all of whom had died of phthisis, with the exception of one sister, who had had gland tumors in the neck and suppuration of the elbow joint. I might add that in these and like cases I took the trouble to verify the patients statements as to the cause of death of their relations, knowing that the less educated of the laity are apt to ascribe a very wide sense to the word " consumption."

Lastly, when all traces of active strumous disease have ceased to be evident, the patient may become phthisical no doubt, but I would urge that that tendency is no more marked in those who were once scrofulous, than it is in those who are simply of enfeebled health. The various aspects of this relationship may be perhaps illustrated by an instance of this kind. Imagine a family of some ten children, the offspring of parents with a tubercular taint. Some of these children become decidedly scrofulous, others remain free from any actual disease, and are simply delicate. Now the former would be less liable to phthisis than would other individuals, while the latter would be especially prone to become consumptive. One would exhibit a negative, the other a positive tendency to phthisical disease.

Some absurd objections have been raised to this identity of scrofula and phthisis that may here be alluded to. Some of the older writers maintain a difference because scrofula is less fatal than phthisis; others because scrofula appears at an earlier age than is common for phthisis, and some few lay stress upon clinical differences. Scrofula tends to appear in early life on account of the unusual activity af the lymphatic system at that period, and phthisis somewhat later, at a time indeed when the lungs are in more active use, when sedentary and perhaps unhealthy pursuits are exchanged for the liberty of childhood, when the modifying influences of puberty are active, and the structural responsibilities of adult life press heavily on an organism never other perhaps than frail. As to clinical differences over and above those already referred to, what arguments can be founded upon them? It has actually been argued that scrofula and phthisis are not identical because the range of temperature is different in the two affections, because the prognosis is less grave in scrofula than it is in phthisis, and because the phthisical waste and become anæmic and sweat at night! Those who support these arguments would be in a position to maintain that acute bubo, acute orchitis, and acute pneumonia are all due to different morbid processes, because their respective symptoms are unlike, their lines of temperature not the same, and the prognosis in each case not identical. It is needless to point out the fallacy of such reasoning; and *à propos* of this last comparison I would, on the contrary, assert that scrofula and phthisis are as much manifestations of the same morbid change as acute bubo, acute orchitis, and acute pneumonia are outcomes of one single process— acute inflammation.

CHAPTER VII.

SCROFULA AND ACUTE MILIARY TUBERCULOSIS.

Acute miliary tuberuluosis should be kept distinct from those other diseases generally described as tuberculous, just as pyæmia, with its multiple abcesses, may be kept apart from common suppuration. Acute miliary tuberculosis is an infective disease, a disease due to the dissemination throughout the body of some noxious material, the nature of which is not yet fully known. It would appear, however, that some pre-existing caseous mass may provide the infecting agent, and according to Buhl, Leudet, and others, such caseous masses are seldom absent from the cadaver after death from this disease. It might fairly be compared to pyæmia, the only difference between the two being the nature of the poison and the special local evidences it produces—on the one hand, an eruption of abscesses, on the other, an eruption of tubercles. As scrofula is an affection that leads almost constantly to caseous products, it is no matter of wonder that scrofula has been considered as a primary cause of acute miliary tubercluosis.

It must be understood, however, that scrofula acts only by producing a caseous material, and that all other conditions that lead to caseation may be regarded equally as causes of acute miliary tubercluosis. Why that material becomes absorbed in some cases and furnishes a noxious infecting agent, while in others it remains harmless, cannot yet be determined. A vast multitude of individuals must pass the greater part of their lives with deposits of cheesy matter in their bodies, and yet not become the subjects of general tuberculosis.

Indeed, the great frequency of caseous matter in the tissues, and the comparative rarity of acute tuberculosis, detracts very considerably from the grave signification that is supposed by some to attach to the presence of such matter in the body.

So far as scrofula itself is concerned there is no relation between the severity of the disease, its situation, the extent of its cheesy products, or its effect on the general health, and the liability to general tuberculosis. This disease has been traced to caseous glands, to scrofulous orchitis, to tubercular disease of the bladder, to ulcers of mucous membranes, to scrofulous caries, to inspissated pus, to the residues of other than strumous inflammations, to phthisis, and indeed to every condition that may involve caseation. Viewed from a practical point of view I do not imagine that the prospect of acute miliary tuberculosis in scrofula can effect either in one way or in another any mode of treatment that may be proposed. It neither makes us, on the one hand, extremely anxious to remove at once from the body every caseous tissue that operation can remove; nor does it leave us, on the other, absolutely callous as to the future of a patient who retains caseous deposits in his body.

CHAPTER VIII.

THE ETIOLOGY OF SCROFULA.

Scrofula is a disease that may be both hereditary and acquired.

With regard, in the first place, to heredity, a tubercular parent hands down to the offspring that particular phase of tuberculosis known as scrofula. The parent may present any form of tuberculosis, and may be scrofulous or phthisical, or the subject of any of those diseases commonly classed as the tubercular.

It is only when such transmission occurs that scrofula can be strictly regarded as an hereditary affection; for by heredity in disease one assumes that the malady in the offspring is identical with that in the parent, or is at least no less than a modification of it. The question as to the probability of any condition of ill-health in the parent, other than that due to tubercular influence, causing scrofula in the progeny, will be considered by-and-bye. It is here only needful to remark that a vast number of the most diverse diseases in the parents have been considered as active in that direction.

Phthisis in the parents is an extremely common cause of scrofula in children. Lugol* asserts that more than one-half of all scrofulous patients have had phthisical progenitors. Out of 141 cases of scrofula investigated by Balman, in 9 instances the father had died of phthisis, and in 11 the mother; while among the near or distant relations of these scrofulous patients 67 deaths from phthisis had occurred on the mother's side, and 89 on the father's side. I made a detailed investigation into the family history of 65 scrofulous patients—having especial reference to this matter of phthisis—with the following results. In 27 of these cases I could find no trace of phthisis among any members of the patient's family either near or distant. In 13 instances the father had been phthisical and in 6 instances the

---

* "Loc. cit.," p. 46.

mother. In the remaining 19 cases both father and mother were free from phthisis, but in 9 of these instances deaths from phthisis had occurred among the mother's relations, and in 10 among the father's relations. The patients, whose family history was the subject of this inquiry, presented scrofulous disease under different aspects, but the majority of them were suffering from the glandular form of scrofula. It will be seen from these cases that phthisis in the father is a potent cause of scrofula. In several instances where this condition obtained the form of struma was markedly severe. The influence of this ill-health in the father is often very conspicuous in cases where the phthisis has not appeared until after several children have been born, and where the mother is healthy. As the point is important, I will cite two cases illustrative of this.

A woman, aged 47, had 13 children. She herself always enjoyed perfect health, was vigorous and robust. and there was no suspicion of scrofula or phthisis in any branch of her family. Her husband died at the age of 45 of phthisis, which disease had first shown itself some five years before his death. Of the 13 children, 3 died before the age of $1\frac{1}{2}$ years from, respectively, acute bronchitis, scarlet fever, and convulsions. Four of the children were born during the last six years of the husband's life, and these are all, without exception, scrofulous. The remaining 6 children have perfect health, and have shown no traces of struma. The family were in good circumstances, and the mother could in no way account for the delicate health of her younger children. A similar case was kindly communicated to me by Dr. King Kerr, of Leytonstone. The father died at the age of 37 of phthisis, having been phthisical for three years. The mother—who was of the same age as her husband—was a perfectly healthy woman. Five children were the result of the marriage. The first two children were boys, and are free from any trace of scrofula; the next two children died in infancy of simple ailments; the fifth child was born two months before the father's death, *i. e.* the father was advanced in phthisis when the mother was impregnated. This infant, at two months old, had impetigo of the scalp, followed by enlarged glands in the neck. When ten months old the glands in the neck suppurated. Enlarged glands then appeared in the groin and axillæ, abscesses formed in many parts, the mesenteric glands became affected, and the child died at the age of one year of tubercular meningitis.

These and like cases appear to show the potency of phthisis in the father as a cause of scrofula; and *à propos* of this subject it may be observed that certain writers have pointed out that in all cases of scrofula the health of the father has exercised a more deleterious effect in the causation than has the health of the mother.* This statement is, however, too wide and general to permit of its being accepted.

Scrofula in the parents is another common cause of scrofula in children, although it is a less frequent factor in the etiology of the disease than is phthisis. It is singular also that in the majority of cases the scrofula is in the mother.† At least such is the result of my own observations. There is no uniformity in the disease transmitted. The mother may have suffered from gland disease and the child may have a strumous joint, or spinal caries, or lupus; or the parent may have had any of these affections, and the child have glandular lesions. It is interesting, moreover, to note how interchangeable are scrofulous and phthisical influences. Phthisis in one generation may appear as scrofula in the next, and perhaps as phthisis again in the third. Atavism is observed in the transmission of these diseases, although, I believe, not very frequently. I have notes of a case where the grandmother died of phthisis, the mother had sound health, and her child was scrofulous. Tyler Smith‡ insists on the possibility of one generation being entirely

---

* See, for example, "De l'Adenopathie cervicale chez les Scrofuleux. These de Paris." No. 469, 1879, by Dr. L. Deligny, p. 28.

† Phillips ("loc. sit.," p. 119) found, on examining indiscriminately a large number of parents, that when the father was scrofulous and the mother sound, the children were strumous in 23 per cent. of the cases examined; and in instances where the mother alone was scrofulous, 24 per cent. of the offspring showed evidences of scrofulous disease.

‡ "Loc. cit.," p 12.

missed over in this manner, and Hueter* expresses an identical opinion.

The manner in which the disease is distributed over the members of a family in cases of heredity is often incomprehensibly irregular. One child alone may be scrofulous out of a family of six or eight, and no reason found to account for the selection. Some of the children may be scrofulous, some phthisical, and the rest simply delicate. Sometimes the younger children are more scrofulous than the elder; or, on the other hand, the elder children may be severely affected, and the succeeding progeny but slightly influenced by the disease. And these inequalities in distribution may be independent of any modification in the health or circumstances of the parents, of any difference in diet, of any change in hygienic surroundings, and appear to be so far really unaccountable. In some cases, however, where a scrofulous or phthisical taint already exists the effects of mere ill-health in the parents is strikingly shown in the children. Dr. Kennedy† gives a good illustration of this. A peasant, with a history of scrofula in some members of his family, married a healthy woman. Two children were born, who remained quite sound and well. The man was then attacked with rheumatic fever, and subsequently endured great poverty. During this period of depression two other children were born, both of whom became scrofulous. Finally, the man regained his health, and had other children, who were as healthy as the first. (One must assume in this instance—an assumption all would not allow—that the scrofula in the two children was not wholly of the acquired variety.)

I would therefore quite disagree with the dogma of Lugol, who says that "if there be one fact in pathology more impossible than another, it is that one child should be scrofulous and his brothers and sisters perfectly free from the taint."‡

---

* "Loc. cit,," p. 264.
† "Natural Selection in Scrofula." *Brit. Med. Journ.* vol. i. 1874. p. 252.
‡ "Loc. cit.," p. 19.

I have met with scrofulous children who have had brothers and sisters in whom, I think, the most suspicious could detect no evidence of any "scrofulous taint," unless one allows the term scrofula to include the enormous area now occupied by the simply delicate and the non-robust. Lugol readily disposes of all exceptions to his rule by assuming adulterous intercourse on the part of the mother.

Now comes the question—Can conditions of ill-health in the parents other than those due to tubercular influence cause scrofula in the children?

Sir J. Clark states, *à propos* of this matter, that "a deteriorated state of health in the parent, from any cause, to a degree sufficient to produce a state of cachexia, may give rise to the scrofulous constitution in the offspring."* Many subsequent writers have endorsed this statement, but I am inclined to think that it has been somewhat too widely accepted. A fair number of cases of scrofula are met with in patients whose parents show neither trace of scrofula nor tendency to phthisis, but who are simply in a "deteriorated state of health." A detailed examination of such cases gives in many instances some such results as these: either the parent had some manifestation of scrofula in youth, all traces of which have since disappeared, or there is a history of scrofula or of phthisis in some member of the family, near or distant. As instances of this latter condition I have seen scrofula in children whose parents were apparently healthy, but who had an aunt or a cousin the subject of strumous disease. Had one or other of these parents been in a cachectic condition, that cachexia would have been ascribed as the sole cause of the scrofula in the children, if Sir J. Clark's axiom were upheld.

In one case that I examined, the father and mother were both free from actual disease, although the mother was delicate, and there was no history of scrofula or phthisis in any of their relations. Their youngest child was, however, scrofulous. Excluding acquired scrofula,

---

* "A Treatise on Pulmonary Consumption," London, 1835, p. 222.

the case appeared to support the above axiom, until in about twelve months' time the mother developed pulmonary consumption.

Still, there are instances where such explanations do not hold good, and where one cannot avoid recognizing some defects in the parents' health other than those due to either scrofula or tubercle. I do not, however, think that these influences are common, as an inquiry into various cachectic conditions will show. Cancer, chronic lung disease (other than phthisis), and chronic kidney disease, may produce a "state of cachexia," and they are all common affections; yet we by no means find that scrofula is unduly common in the children of such individuals. Observation would, indeed, make one very loath to regard these diseases as causes of struma in an offspring. Dr. Grancher applies a similar observation to "the whole race of chlorotic, dyspeptic, and cachectic persons," and is not disposed to believe that they can transmit to their children the tendency to ill-health known as scrofula.

I think it is to syphilis we must turn for an example of this present matter, and it is indeed probably the only disease other than tuberculosis that can readily or even occasionally produce scrofula in the offspring. I have, I think, undoubted proof in several instances that syphilitic parents may beget strumous children. One or two of such instances were of this character: the parents had been syphilitic, the elder children had presented syphilitic symptoms, such as interstitial keratitis or stomatitis, leading to deformed teeth, while the younger children were simply scrofulous, and came with eczema, phlyctœnular ophthalmia, enlarged glands, &c. In less modern times this relationship between scrofula and syphilis was considered to be very close, inasmuch as the manifestations of hereditary syphilis were included under the head of scrofula. It is extremely difficult to say under what circumstances transmission of the disease occurs; and even those who strongly maintain the connection of the two disorders do not venture any suggestions upon this point.

Among other causes reputed to produce scrofula in the progeny are advanced age of the parents, disproportion in their respective ages, and especially advanced years in the father; marriage of near kin; and the usual scapegoats for all obscure influences—intemperance and sexual excess. It must be confessed that the effect of these supposed causes has not been as yet demonstrated.

Among the general predisposing causes of scrofula may be mentioned—

1. *Locality and Climate.*—It is stated that scrofula is much more common in some regions of the world than in others, and that it is more common in the temperate zone than in the extreme north or in the tropics. The statements of authors upon this head are, however, most contradictory, and it is evident that a sufficient number of well-authenticated statistics are not at present forthcoming on which to establish any conclusions as to this point. Some writers have gone to great extremes upon this subject, and according to Henning* scrofula is nothing but a climate disease, and is due to certain atmospheric influences. It is possible, however, to understand that scrofula is likely to be more prevalent in cold and damp districts than in warmer and drier climates, if for no other reason than this—that cold and damp, and the greater confinement within dwellings necessitated by a cold climate, would be apt to produce those catarrhs that are known to be so often the exciting causes of scrofula. The matter, however, does not rest here, for the habits of the various races must be considered, as well as their modes of living, their diet, and their general hygienic surroundings.

2. Certain months of the year and certain seasons have been considered to effect the production of scrofula. Phillips and Lugol assert that the spring is the season in which scrofula most commonly makes its appearance. Tyler Smith cites the months of April,

---

* "A Critical Inquiry into the Pathology of Scrofula," by S. G. Henning, M.D., 1815, p. 107.

May, October, and November as the periods most concerned in this matter. As these statements refer mainly to the production of glandular disease, they are quite intelligible when one recollects how frequently gland disease depends upon a mucous catarrh, and how such a catarrh acquired in the winter is likely to lead to a scrofulous manifestation in the spring, or how the asmospheric conditions of spring itself may not be inactive in this direction. Probably, therefore, these statements, so far as they refer to gland disease, are to some extent correct.

3. Age.—Scrofula is essentially a disease of early life. The marked implication of the absorbent system in scrofula, on the one hand, and the great activity of that system in early life, on the other, are well-nigh sufficient to explain this fact. The glands in young children are comparatively larger than in adults, while the more conspicuous masses of adenoid tissue in the body, such as the tonsils, Peyer's patches, and the solitary glands are also unduly prominent. In perfectly healthy children, under the age of three, and who are not too stout, I have often been able to feel glands in the posterior triangle of the neck that are not obvious to the touch in even very thin adults. As age advances, the absorbent system becomes less active, and in the old the glands are often shriveled and hard, and much smaller than are those met with in the prime of life.

As we have seen how constantly the lymphatic tissues are implicated in scrofula, it will follow that these facts explain not only the undue frequency of glandular disease in the young, bnt the occurrence also of bone and joint affections, of ulcers, of cold abscess, and other strumous manifestations. There is a fair amount of uniformity among authors as to the commonest time of life for the appearance of scrofula. Thus Lombard gives from four to eight years old. Balman's statistics show that 73.76 per cent. of the cases of gland scrofula occurred between the ages of two to fifteen years. Tyler Smith asserts that two-thirds of all gland cases occur before twelve years of age. Hueter gives the commonest periods for the commencement of scrofula as between three and twelve years, and Birch-Hirschfeld as between three and fifteen.

From an analysis of 509 cases in the records of the Margate Infirmary, I find that the largest number of scrofulous disorders of all kinds have made their first appearance between the ages of five and fourteen. The superficial affections of the skin and mucous membrades are not included in these statistics. They undoubtedly occur at a still earlier period, and are among the very first manifestations of scrofula. Owing, moreover, to restrictions as to age at the Infirmary, these statistics do not include all those who may have died or may have been cured before the age of six; and, allowing this, it is probable that the ages given by Birch-Hirschfeld of three to fifteen are more correct than those resulting from these records.

An analysis of the Margate cases in this matter of age will be found on the next page.

Another period when scrofula not unfrequently appears, is between 20 and 30, and then a little after 30. Rindfleisch mentions the former period as one common for the development of hereditary scrofula. So far as my own experience goes, such disease is more common in females, and is apt to present itself in the glandular apparatus. In three cases of adult scrofula reported in the "Transactions" of the Pathological Society, and already referred to, the ages of the patients were respectively 45, 33, and 30, when the disease commenced. They were all females, and the disease in each case glandular. Other cases will be alluded to on the chapter on Gland Affections. Lastly, scrofula may appear—perhaps for the first time—in old age. Sir James Paget* was the first to fully describe this condition under the title of "senile scofula," and the subject has since been very exhaustively treated by Dr. Bourdelais.† The matter will be alluded

---

* "Clinical Lectures and Essays." London, 1875.

† "Sur quelques Observations de Scrofule chez le Vieillard. These de Paris," No. 297, 1876. See also "Considerations sur quelques Affections scrofuleuses chez le Vieillard," by Dr. Dumoulin. Paris, 1854.

THE ETIOLOGY OF SCROFULA.

TABLE

Gland Disease. Males. Total 66.

Ages at which the disease commenced	1-4	5-9	10-14	15-19	20-24	25-29	30-34	35-39	40-44	45-49	Not known
Number of patients in each division of age	10	21	14	6	1	3	1	—	1	—	9

Gland Disease. Females. Total 89.

| Number of patients in each division of age | 8 | 22 | 34 | 9 | 5 | 2 | 1 | 2 | 1 | — | 5 |

Bone Disease. Males. Total 75.

| Number of patients in each division of age | 1 | 18 | 18 | 10 | 8 | 7 | 1 | — | 2 | — | 10 |

Bone Disease. Females. Total 77.

| Number of patients in each division of age | 8 | 20 | 23 | 9 | 7 | — | 1 | 1 | 1 | — | 7 |

Joint Disease. Males. Total 86.

| Number of patients in each division of age | 6 | 28 | 18 | 13 | 5 | 4 | 1 | — | — | 1 | 10 |

Joint Disease. Females. Total 57.

| Number of patients in each division of age | 6 | 8 | 22 | 11 | 1 | 1 | 1 | 3 | — | 1 | 3 |

to subsequently. In the 16 cases reported by Dr. Bourdelais, the ages of the patients are as follow: one patient was 41 years of age, and another 57; five were between 60 and 70; seven were between 70 and 80; and two were more than 80 years old.

Scrofula is comparatively uncommon before the age of one year. Such cases as I have seen under that age have mostly been severe, and have in every instance been associated with distinct heredity. Most of the cases of "infantile struma" described by the older authors are evidently cases of hereditary syphilis.

4. *Sex.*—With regard to sex, I believe no distinctions are observed in scrofula. It is equally common both in male and female. Reliable statistics on this head are almost impossible to obtain, owing to the fact that while

severe cases of the disease come under hospital or dispensary treatment, the milder cases, the majority, are met with in comparatively small numbers. The 509 Margate cases are thus distributed :—

	Males.	Females.
Gland disease	66	89
Bone disease	75	77
Joint disease	86	57
Other cases	21	38
Total	248	261

These figures, however, cannot be regarded as of much value, beyond showing, perhaps, that gland disease appears to be more common in the female, and that joint affections are more common in the male, due possibly to the greater liability to injury in that sex. The researches of Dr. Bourdelais show that senile struma is more common in the female, and I think that that remark may apply to most cases of adult scrofula.

5. *Complexion.*—There appears to be no possible connection between scrofula and any particular complexion. At one time it was stated that scrofula occurred in the fair-haired, and a mass of figures was brought forward to substantiate this fact. It was then discovered that it was more common in the dark-complexioned, and an equally large mass of figures supported that observation also. Conclusions on this point are absolutely valueless, unless the observer can give, at the same time, the general proportions of fair-and dark-complexioned persons in the particular country or locality in which his investigations on scrofula have taken place. In the cases I obtained from the Margate records, details as to complexion, color of hair and eyes, &c., are given in 450 cases. Of this number 221 were air, 115 I classed as "medium," and 114 as dark. From these statistics one can only conclude that in England," or at least in that class of society from whom these scrofulous patients are drawn, the fair-complexioned are in the majority. Phillips's investigations show only 32 per cent. as presenting light hair and eyes. In 49 cases at l'Hôpital

de Berck, recorded by Dr. Deligny,* 3 had brown hair, 28 a deep chestnut, 11 light chestnut, and 7 blonde.

Acquired Scrofula.—Although I would strongly urge that in the great majority of all scrofulous cases a tendency that favors the particular process of scrofula has been inherited from the parents, it must be owned that in some instances the disease may be independent of such heredity, and be developed *de novo*. I believe cases of pure acquired scrofnla to be uncommon, and certainly less frequent than in former days. The circumstances under which scrofula may be acquired are very numerous, and for the most part imply simply such influences as would lead to general bad health. As the most important, may be mentioned—bad ventilation and overcrowding; absence of sunlight; insufficient, bad, or unsuitable food; cold and damp; imperfect clothing; and, indeed, all those conditions that are the common surroundings of squalor and poverty. To scrofula developed under many of these circumstances, one might well apply the term of Grancher, " la scrofule a miseriâ." It may thus be said, therefore, that acquired struma is practically limited to the poor; and continuing the same argument, it may reasonably be assumed that the great majority of cases of scrofula among the rich are due to heredity. Upon this latter point M. Chauffard and others have especially insisted.† I think general experience will, however, allow that scrofula—especially its severer forms—is much less common among the rich than among the poor. Apart from the many advantages of the wealthy, the poor, while most prone to develop acquired forms of struma, are not exempt from the hereditary phases of the disease. As many of the conditions just referred to as inducive of scrofula are more apt to be met with in crowded habitations, it happens that scrofula is somewhat more frequent in large towns than in the open country. The disproportion would appear, however, to be slight, for although the child of a farm

---

* "Loc. cit.," p. 25.
† M. Chauffard. "Premiere livraison du Correspondant," July 1870, p. 172. See also MM. Perrochaud and Cazin, " Soc. de Chir.," April 1876.

laborer may have plenty of air and light in the daytime, it is as likely to be improperly and imperfectly fed as is the town child, and probably occupies an apartment at night that in the matter of foul air and filth could not be well beaten in the purlieus of the dirtiest city. As examples of acquired scrofuia I may cite the following :—

A girl, aged 16, had extensive gland enlargements on both sides of the neck. These had been noticed 8 months. The patient was the fifth child out of a family of eleven. Her father and mother were healthy. There was no trace of phthisis or scrofula in any branch of the family, nor in any of the other children. Five of the eleven had died in infancy of simple ailments. These were respectively the first, fourth, seventh, ninth, and eleventh children. For the last 18 months the patient had worked in a small close room, had had little or no outdoor exercise, and been very indifferently fed. Previous to this time the family had for some months experienced severe poverty. As compared with the rest of the family the child had always somewhat been delicate. To take a case later in life. A female, aged 20, had greatly enlarged glands in the neck, which had existed for 14 months, and had suppurated. Her parents were healthy. There was no phthisis in any of the family, and all her brothers and sisters were free from any trace of scrofula. Some year or so before the gland disease appeared, she began a very sedentary employment, and worked in a close, ill-ventilated room. As she endeavored at the same time to support herself, her food was poor and insufficient.

The best examples, however, of acquired scrofula have been furnished by poor-houses and prisons. The facts from these sources are very clear. Patients who have always enjoyed good health, who have no trace of tubercular mischief in any member of their family, enter one of these of institutions. Close confinement, want of exercise, poor and insufficient food, and perhaps plenty of work, soon take effect, and many under these conditions develop distinct scrofulous disease.

It must be said, however, that examples of scrofula

from these sources are much rarer now than they were formerly, owing to the greatly improved hygienic conditions of workhouse inmates and prisoners at the present time. As an instance of "parochial struma," I might quote this observation from Tyler Smith's book.* In a worknouse in Kent there were on April 29, 1841, 78 boys and 94 girls. It is stated that all these children were healthy when admitted and free from scrofula. When examined enlarged glands were found in all the boys and in 91 of the girls. The following was the diet these children enjoyed—bread and cheese for dinner four days in the week, suet puddings and vegetables on two days, and meat (in small quantity) on the remaining day of the seven. One of the best accounts of what Autenrieth † calls "penitentiary scrofula," is given by Dr. Deligny in his statistics of the disease in certain French prisons. He gives a vast number of instances of prisoners from 25 to 40 years of age or older, who entered the prison without a trace of struma, and yet developed gland tumors that suppurated before a long period of confinement had passed. Indeed, in six instances, scrofula appeared within eight months of admission. His statistics extend from the year 1861 to 1872, and he ascribes the occurrence of the disease to bad air, want of exercise, and poor and insufficient food. On remedying these evils the patients at once improved. Dr. Grancher‡ remarks that after the siege of Paris he met with many persons who had become anæmic, and had developed gland tumors. They recovered, but in many cases the masses suppurated, and some became caseous.

The actual *exciting causes* of scrofula are very numerous. A tendency to scrofulous inflammation being inherited or acquired, it needs but a slight exciting cause to produce an outburst of the disease. Many of these exciting causes will be referred to more in detail when treating of gland disease. Speaking generally,

---

* "Loc cit.," p. 39.
† "Spec. Nosologie und Therapie." Wurzburg, 1836, ii. p, 333
‡ "Loc. cit,, Dict. Encyclop.," p. 343.

one might say say that any condition that impairs the health the health of a person predisposed to scrofula is sufficient to bring about some manifestation of the malady. Thus scrofula often appears for the first time after the eruptive fevers. Measles appears to be very commonly an arouser of the scrofulous process. It acts, like scarlet fever, not only by temporarily impairing the health, but also through the catarrh that is a symptom of both diseases. Scrofulous manifestations have sometimes appeared for the first time after vaccination. This might be merely a coincidence, or might be due to the slight impairment of health, associated with vaccination, acting upon a subject already predisposed to struma. Richard Carmichael* considered digestive disorders as the usual exciting cause of scrofula, and others have endorsed this view. The statement, I believe to be fallacious, except perhaps in a very few instances. The digestive disorders of such patients are, as a rule, actually due to a scrofulous affection of the mucous membranes, and the disease therefore is already existing, however caused. In some cases scrofula has appeared for the first time during pregnancy or lactation, and in cases where the disease has existed in childhood these conditions often cause it to reappear.

In concluding the subject of etiology I must express a belief that scrofula is on the decrease, and that the manifestations of the disease are, on the whole, not so severe as they were in former days. I am aware that many of the older authors described as scrofulous cases which we now know to be syphilitic, or even cancerous, or at least due to a disease other than that with which we are now concerned. But making all allowance for this, the account in the past of the prevalence of scrofula could not apply to the present time, We read of workhouses, schools, and penitentiaries in the early part of this century the inmates of which were nearly all scrofulous. We read of scrofula as a wide-spread, severe, and distressing disease among the inhabitants of little towns and villages where now it is but little

---

* "On the Nature of Scrofula," 1810, p. 20

seen.* We hear of immense gland enlargements leading to suppuration and death as of no uncommon occurrence; and making every allowance for errors in diagnosis and a reckless use of the word scrofula, the mortality from the disease is certainly much less than it was years ago. Common as scrofula was in Carmichael's days, he speaks of it as "more common years ago;"† and Phillips, who was greatly impressed with the amount of struma in England in 1846, concludes that " scrofula is much less prevalent at the present day than it was in the seventeenth and eighteenth centuries." ‡ And coming more to detail, Mr. Vernon, § speaking of strumous eye affections, maintains that we rarely meet nowadays with such severe cases as were common in times gone by.

The dimunution of scrofulous diseases is easily to be explained by the vast improvements that have been made of late in sanitary science, by the better condition of the poor as regards all matters of health, and by advances possibly in the science of medicine and in the treatment of disease.

CHAPTER IX.

THE SCROFULOUS INDIVIDUAL.

THE physiognomy of scrofula, the type of face and form supposed to be indicative of the disease, have for ages been subjects upon which writers have loved to exercise their imaginative and descriptive, powers. Some extraordinary pictures have been given of the scrofulous individual, who has at one time appeared repulsive and at another peculiarly pretty. Some of the

---

* See Hamilton's observations, for example, on the amount of scrofula in the town of Lynn in 1781, *loc cit.*, p. 160.
† "Loc. cit.," p. 54.
‡ "Loc. cit.," p. 98.
§ "Scrofulous Iritis," *British Med. Journ.*, vol. ii, 1874, p. 276.

older writers describe the physiognomy of the strumous with such precision that had their statements been even partially true, all sufferers from that disease could have been recognised at a glance. Other observers divided the scrofulous into two classes—the sanguine and the melancholic, each with well-marked and distinctive features and appearance. This division, while ingenious and affording great opportunities for the exercise of fancy, had the practical disadvantage that the bulk of scrofulous persons belonged neither to the one class nor to the other. Writers in later times have possibly gone to the other extreme, and have asserted that there is no type of countenance or figure that is peculiar to the strumous, or that is even of common occurrence among them. Others, again, like Bazin,* have so generously described "the scrofulous habit," that it might include every individual who was not conspicuously robust.

Many of these discrepancies depend upon incorrect ideas of the clinical characters of scrofula, and upon the forced attempt of the older pathologists to associate every disorder with some distinctive physiognomy. Several of the older descriptions of the scrofulous face or habit are compounded from the features of phthisis, rickets, and hereditary syphilis, ; while not a few simply coincide with the outward manifestations of one or other of these diseases.† Bredow had never seen a hare-lip but in a scrofulous child, and Macartney discovered certain mental features in the scrofulous "which more strikinkly indicate the peculiar state of the constitution than do all the other signs." Moreover, many of the characteristics of the strumous physiognomy, as described by some, belong in reality to the already developed disease, and among such features may be mentioned the swollen and thick neck, the enlarged upper lip, the sore eyes, the tumid eyelids. Omitting further discussion as to opinions that have been held, and reviewing the mat-

---

* " Lecons theoriques et cliniques sur la Scrofule," 2nd ed. 1861. Paris.

† See, for example, "The Pathology and Treatment of Scrofula," by Robert Glover, M. D., 1846, p. 145.

ter from the clinical and pathological bases I have already given, it may, I think, be said that there is no physiognomy quite distinctive of scrofula, no type of face or form so peculiar to the disease, or of so common occurrenc among its victims, that it is possible to recognise in all cases the "scrofulous habit" considered apart from actual manifestations of struma. Scrofula may occur in persons of perfectly healthy aspect. I can call to mind a little lad some ten years of age who came to me with gland masses in the neck, and whose rosy and chubby cheeks and general bearing of robust health singled him out from among a number of less vigorous out-patients. Speaking generally, the physiognomy of of scrofula is the physiognomy of poor health. The most that can be said of the aspect of many strumous patients is simply that they do not look well, that they are delicate in appearance. Our scanty knowledge of the factors of simple frailty of health scarcely enables us to say more than this of many strumous paients. Put a number of such scrofulous children together, and one can merely say that they look out of health. There is perhaps no conformation of face, no particular features, common to even the majority of them. When these children present glandular swellings or chronic joint diseases or certain bone affections, it is then easy enough to say that they are scrofulous; but looking at them without any known manifestations of scrofula at all, who will be bold enough to call even a few of them "scrofulous," or distinguish them from a mass of children that can be seen crawling about the slums of a great city, and that merit no higher scientific term than that of being "seedy-looking"?

Excluding scrofulous individuals that on the one hand, look robust, and on the other, merely out of health, we at last arrive at a class of the strumous who present something approaching distinctiveness in their physiognomics.

The general features of this class are sufficiently well marked to enable us to separate them into two divisions, that, for the want of better words, may be known by the old terms—the *sanguine* and the *phlegmatic* types of scrofula. It would, I think, be well to speak of these as types rather of defective health than as types of that special form of ill-health known as scrofula; for while persons showing the features of one or other of these classes are for the most part scrofulous, the whole are not. And these exceptions—very few although they may be—render it impossible for us to say that every individual presenting the characters of one or other of these typical classes *must* be scrofulous. If but a few non-syphilitic children were found with "Hutchinson's teeth," depressed noses, and prominent brows, the physiognomy of hereditary syphilis, would cease to be typical, and would occupy the position with regard to syphilis that I believe the "sanguine" and "phlegmatic" types of scrofula occupy to that disease.

It will now be convenient to describe these types of unhealthy person that are so frequently met with among the strumous.

1. The *sanguine type*. Individuals placed in this class are credited with these features, and they refer more particularly to children. They are tall, slight, and graceful, with well-formed limbs, hands, and feet, a fine clear skin, and usually a fair complexion. The face is oval, the lower jaw small, the features delicate and regular, the lips thin. The eyes are bright, and covered with long lashes, and the hair is often remarkably fine and silken. A sprightly and excitable disposition may be added, and the picture is complete. These features are identical with those described by Sir W. Jenner* as typical of the tubercular child, and they also very fairly accord with the usual type of the phthisical individual.†

The leading points of this physiognomy were admirably shown in a series of photographs exhibited in the Museum of the International Medical Congress by Dr. Mahomed and Mr. Galton.‡ By some special process a

---

* Article on "Tuberculosis," *Med. Times and Gaz.*, vol. ii. 1863, p. 423. See also Art. by Dr. Laycock, on "The Physiognomical Diagnosis of Disease," *Med Times and Gaz.*, vol. i. 1862, p. 341

† See description by Ruehle, "loc. cit.," p. 510.

‡ Some details are given in the "Museum Catalogue," 1881, p. 81. Also in "Abstracts," p. 152,

"composite" photograph is produced of many individuals. In this composite picture—a single face—"all that is common remains, all that is individual disappears." The typical or average face thus produced from a number of phthisical women agrees with the description above given, and is a face, it must be owned, that is singularly graceful and delicate. Some physicians have described a "catarrhal diathesis," a tendency to frequent inflammation of mucous membranes; and the features ascribed to persons with such a diathesis accord also with this so-called sanguine form of scrofula.* It is obvious that under various terms and in connection with various diseases, the same type of unhealthy individual has been described. Lastly, the name of serous or erethic scrofula has been given by this type.

2. In the phlegmatic type are comprised individuals as a rule short and burly, with coarse limbs, large hands and feet. The face is broad, the lower jaw large, the malar bones often prominent, the features coarse and irregular. The nose is generally thick, the lips tumid, the lobes of the ears large, and the neck unshapely. The skin is coarse, harsh, and thick. The amount of subcutaneous cellular tissue is considerable, and often sufficient to conceal the muscular outlines of the body. The skin in the previous type is fine, and it is possible to pinch up with the fingers a little portion of it; but in these individuals none but a large fold of skin can be picked up, as it is so coarse. Speaking generally, persons of this class appear flabby and heavy-looking; they are apathetic, have little muscular power, and are soon tired. The vascularity of their tissues appears to be impaired, and leads to certain peculiarities of parts that will be dealt with subsequently. This type is very well represented in the photographic series of Dr. Mahomed and Mr. Galton, under the title of "coarse struma," the examples being all obtained from phthisical patients. Older authors described a like individual under the name of melancholic scrofula, and many

---

* "De la Diathese catarrhale des jeunes Filles," by Dr. Richelot. "L'Union Medicale," March 3, 1881, p. 367.

accounts of what is known as the lymphatic temperament accord with the above.

It must be distinctly understood that these descriptions are merely typical. Many scrofulous individuals—as before stated—can perhaps not be placed with certainty either in the one division or in the other. Moreover, out of the whole mass of the strumous there may be comparatively few who would present all the features of one or other of these types; and those who expect to find commonly among the scrofulous physiognomies so marked as those above detailed will certainly be disappointed. Apart from all this, however, it is possible to class a vast number of the subjects of struma according to the types I have described. The test applied to effect this classification need not be elaborate. If with one aspect of scrofula be classed all those with oval faces, regular features, and fine skin, and with the other aspect those with broad faces, coarse features, and thick skin, an approximate result will be obtained which will be of value.

As is common with all hard and fast descriptions of individuals, a large number of the strumous belong to a kind of medium type between the two just given.

Such a type would include what is known as "pretty struma." The general features of individuals, so termed, belong to the so-called "phlegmatic" type; but the coarseness of the features is toned down; the lips would be called "full," not tumid, and a coarse flabbiness would subside into a pretty plump condition of the body. The limbs, if not actually graceful, are at least prettily rounded. The skin may not be thin and fine, but it is soft, white, and clear. The general expression is not absolutely apathetic, but would be termed rather gentle, and eminently feminine. An excellent representation of "pretty struma" was given in the photographic series above alluded to.

The practical aspects of this matter of physiognomy are of no little importance. To the first, or so-called "sanguine"* type belong those cases of scrofula that

---

* I continue to retain these ancient words "sanguine" and "phlegmatic" for want of better. They have the advantage of being almost

show distinct heredity, and especially those that are in some way or another intimately associated with phthisis. Patients in whose family history there is a strong element of phthisis or tuberculosis nearly always present the features of this class. In these individuals the tubercular process appears to reach its greatest development, they are liable to the more severe and fatal forms of the disease, and they offer, in consequence, the elements of a somewhat more unfavorable prognosis. The physiognomy is identical, as before remarked, with that accredited to phthisis and tuberculosis generally; and the few cases that I have seen, where scrofulous subjects have succumbed to pulmonary consumption or general tuberculosis, have been individuals of this class. It must not for a moment be assumed that a tendency to the graver forms of tubercular disease is *peculiar* to such of the strumous as have this physiognomy, but it certainly is more usual among them, and I think that reported cases will bear out this assertion. The scrofula of the rich, which is so generally independent of acquired causes, has usually the features of the "sanguine" type, and the graver prognosis of the disease in such individuals has already been commented upon. The term "phthisical form of scrofula" has been applied to this aspect of the disease when the physiognomy is well marked, and the term is not inappropriate when the associations of this phase of the malady are considered. Henning * considers this phthisical form of scrofula to be more common among women, but I am unable to express any opinion upon this point.

With regard to the "phlegmatic" type of scrofula, it is the type usually assumed in the acquired forms of the disease. It is best seen perhaps in what has been well termed "parochial struma." From descriptions given it is evident that the large amount of scrofula at one time manufactured by penitentiaries and poor-houses was of this character. At the Margate Infirmary, where

---

meaningless with regard to the present subject, and are thus useful as pure terms.
* "Loc. cit.," p. 79

the patients are drawn largely from the poorer classes, this type of scrofula is very commonly to be met with. Patients with these peculiar features are very liable to great gland enlargements and to sluggish affections of mucous surfaces; they show little or no tendency to the more serious forms of tubercular disease, and although they may relapse again and again, are very readily and satisfactorily improved by treatment. This physiognomy is certainly the one most peculiar to scrofula, and deserves of all others the designation scrofulous or strumous. It is the physiognomy recognized by the older writers, and upon it has been founded the most common and familiar description of the disease. Possibly it is less commonly met with now than in years gone by, and its less frequent occurrence would accord with a diminution both in the number and in the grossness of the examples of acquired scrofula.

This matter of physiognomy appears to further illustrate the relationship between scrofula and phthisis upon which I have already insisted. Phthisis is met with in both the types of disease already described, although it is much more commonly associated with the so-called sanguine physiognomy. Phthisical individuals with the "phlegmatic" type of face are spoken of as presenting the "strumous form of phthisis," just as scrofulous persons of the sanguine class have been referred to as displaying the phthisical form of struma. In my examination of the phthisical patients at the Brompton Hospital, above referred to, only one out of the seven who had evidence of scrofula was of the phlegmatic type; all the others possessed more or less the features detailed in the first-mentioned class ol the disease. *A propos* of the same subject, Laycock* and others state that it is difficult, if not often impossible, to distinguish by physiognomy alone scrofulous from tuberculous patients, using the latter term in its common clinical sense.

I will now discuss in detail certain features more or less peculiar to the scrofulous, of which mention has

---

**Med. Times and Gazette*, loc. cit., p. 341.

not yet been made, or which have been merely alluded to *en passant*.

1. *The Circulation in the Scrofulous.*—There does not appear to be anything peculiar about the vascular arrangements in the "phthisical form of scrofula," in those individuals who, for the most part slight and frail, have been described as marking the sanguine type or the disease. But in the coarser type of struma defects in the circulation are often very conspicuous. These defects have been frequently alluded to by various writers, but have been lately especially commented upon by Mr. W. K. Treves, of Margate.* In these coarse, flabby, ungainly children, the pulse is often below the average, soft, and wanting in vigor. The blood appears to stagnate in exposed parts, and thus the cheeks and limbs often assume a bluish or mottled aspect. Especially is this seen about the backs of the hands. The extremities appear swollen, as if from cold, and the skin itself feels chilled and clammy. All these features are exaggerated in the winter and npon exposure, but even during the summer weather some of these children retain a refreshing aspect of chilliness. These patients are particularly liable to chilblains, which often take on a very unhealthy action. Indeed, so frequent is this ailment that it forms a feature in the symptomatology of scrofula ; and I have known children to be troubled with "broken" chilblains for eight or nine months out of the twelve. These defects in the circulation also may possibly explain the frequent catarrhs with which such patients are afflicted, and may account, as Mr. W. K. Treves has suggested, for the unwholesome character sometimes noticed in their wounds. As to the cause and real nature of these circulatory defects it is difficult to speak. I believe them to be consequent upon defects in the absorbent system, and on this point would fully endorse the views of M. Potain.† M. Potain has fully described the condition

---

* "The condition of the Circulation in Scrofula," by W. K. Treves, F.R.C.S. *Lancet*, vol. i. 1871, p. 568. From this article I have derived the main points of the description that follows.

† "Art. Lymphatique (pathologie), Dict. Encyclop. des Sc. Med,," vol. iii. 2nd series, p. 475.

just depicted, and speaks of the general state that leads to it as "lymphatism." From a flaw in the absorbent apparatus, there appears to be an excess of nutritive juices in the tissues, they indeed linger there unabsorbed; as a consequence, a kind of solid œdema is produced, and the parts become flabby and sodden. From this material, that should have been removed, a flimsy connective tissue is developed, and the thickness of the subcutaneous layer of that tissne thereby increased. Such a block in the capillary area, where important blood changes are taking place, may well affect the circulation, and induce a vascular stagnation in the part. With this impeded blood current, some of the most characteristic of the changes above described would be associated. I would regard, therefore, these defects in the circulation of certain scrofulous persons as secondary to some fault in the lymphatic apparatus, and such an explanation would well accord with what I have tried to show is an essential feature of the disease.

2. *Temperature.*—Little is to be said upon this point, and further information is wanted. Dr. Lucien Deligny* states that in more than one hundred distinctly scrofulous children examined by him at l'Hôpital de Berck, there was a lowering of temperature from one-half to one degree. These patients, of course, were free from active inflammatory processes. He concludes, therefore, that in the scrofulous the temperature—apart from all inflammation—is below normal.

3. In connection with this subject of circulation and temperature, it may be observed that acute sthenic inflammations are comparatively rare in the scrofulous, especially in those of the "phlegmatic" type. It has also been asserted that general fevers are not apt to run high in such individuals, but to be on the other hand somewhat unduly protracted. During an epidemic of scarlet fever at the Margate Infirmary for scrofula some years ago I observed many cases that appeared to support his assertion.

---

*" De l'Adenopathic cervicale chez les Scrofuleux." These, No. 469, 1876, p. 25.

4. *Menstruation.*—Some assert that the first appearance of menstruation is delayed in the strumous, and others that puberty appears at an earlier age in such individuals. In thirty-nine scrofulous females reported on by Lebert twelve began to menstruate before sixteen, fifteen during the sixteenth year, and the remaining twelve after that age. I believe that this delay in menstruation is only to be met with in some of those patients who exhibit marked defects in the circulation, and in those also who are suffering from some scrofulous disease, inducing anæmia at the time when the function should be established. Apart from this, I imagine that scrofula has no effect upon the appearance of puberty.

In analysing the records of the Margate Infirmary—which records include accounts of females of all ages—I was struck with the great number who were stated to be suffering, or to have suffered, from dysmenorrhœa. I imagine this number to be much in excess of that that would obtain among a like number of healthy girls and women.*

5. *Intelligence.*—The most varied accounts have been given of the mental condition of the scrofulous. They have been accused of displaying precocious sexual passions, and have been credited with an absence of those impulses. They have been distinguished by the possession of certain faculties by one writer, and by the lack of the same by another. An observer who goes more into detail sagely remarks that in the scrofulous "the imaginative faculty preponderates over the reflective," while another has discovered that the great feature of the strumous mind is a "gentleness of disposition and a refinement and judgment in matters of taste." Without raking up more of the ghastly examples of human error that lie buried in the pages of ancient books, I might allude to a very common statement, repeated over and over again, to the effect that scrofulous children are unduly intelligent and precocious. I believe this to be incorrect. One does meet with precocious

---

* Lugol refers to the frequency of dysmenorrhœa among the strumous, but gives no details.

children among the strumous, but that precocity is by no means peculiar. I think such instances are to be explained in this way. The scrofulous child is the delicate one of the family perhaps, it is petted, has more notice taken of it, and is offered every facility for the development of the points that make up the "precocious infant." Some of the poorest children spend half of their earlier years in one institution or another, and by mixing with older children, and receiving more attention from their elders, soon begin to compare in intelligence with their brothers and sisters, who are perhaps indulging in no more intellectual pursuit than that of crawling from one gutter to another. Moreover, the prettiness of some strumous children attracts more attention to them than the bulk of the sickly would perhaps receive.

I knew a strumous boy, aged ten, who conceived a plan of extracting money from the hospital money-box by an ingenious contrivance framed from a piece of firewood and some plaster. His success was great. It would be unfair to such a lad to ascribe his ill-applied ingenuity to scrofula. He was a natural genius.

6. *Certain peculiar features. Hairiness.*—In young scrofulous children one often observes an amount of close-lying, downy hair upon the forehead, especially about the sides of the forehead. A like condition may often be seen on the arms and upon the back, from the occiput to just below the shoulders. As the child grows up this hair becomes less conspicuous, or disappears. In all the cases I have seen the downy growth was very fair. Dr. Wilshire* was, I believe, the first to call attention to this condition, and he regarded it as quite indicative of scrofula.

The value of this sign, however, cannot be maintained until its absence in healthy children has been demonstrated, and its limitation to scrofula or tuberculosis clearly made out. I have seen a certain amount of downiness of the forehead in young children, in whom there was no reason to suspect scrofula or tubercle.

---

* *Medical Times and Gazette*, April 10, 1847.

Ears.—Dr. Constantine Paul* has drawn attention to certain changes in the ears, after they have been pierced by ear-rings, that he considers to be diagnostic of scrofula. The puncture in these cases very slowly ulcerates, the weight of the ear-rings directs the process downwards, and a linear scar is produced, or the ear-ring may cut its way out, leaving a slit, or instead of a scar a linear aperture may be formed in the lobule. If after the ear-ring has cut its way out the lobule be re-pierced, it may cut its way out again, and this may occur three or four times, a considerable amount of deformity being produced. He gives details of 114 cases. In 96 of these cases the patients presented either scars of scrofula or some distinct manifestations of the disease. In 18 there were no direct evidences of struma. In 51 cases the duration of the ulcerative process was noted, and was found to average 4 years and 2 months. In 74 instances both ears were affected, in 38 one ear only. Age appears to have no influence in producing the scars, and in the bulk of the cases the ears were pierced in infancy or childhood. From all these cases accidents were excluded. He concludes with the observation that every female in whose ears the scar left by piercing is not a simple orifice, but presents instead a slit-like aperture or a linear cicatrix, is the subject of scrofula. Dr. Paul states that these scars and slits in the ears are very common. Since the publication of his paper I have examined a large number of women for these changes in the lobule, and I have found so few examples, that, in London, at least, I apprehend such ears are rare. Possibly on the Continent ear-rings are more frequently worn than in England, and they are certainly often of greater size and weight. In all the examples of these ear changes that have come under my notice the subjects were scrofulous, except in one case where the ear-rings had cut their way out on both sides no less than three times. The woman was 30 years of age, free from any scrofula past

---

* " Un nouveau Signe de la Scrofule fourni par les Boucles d'Oreille." *L' Union Medicale* February, 1881, p. 337. et seq.

or present, but very cachectic from syphilis acquired years ago. There was a vague history of phthisis in her family. Before accepting Dr. Paul's very positive assertion, I think these questions should be answered; If the condition is due to general tissue defects, how comes it that in 38 out of 112 instances only *one* ear was affected? How are the 18 cases to be explained where, as Dr. Paul allows, there were no traces of scrofula? and lastly, is there no connection between the lesion and the weight of the ear-ring and the metal of which it is composed? I must, moreover, state that in many of the cases detailed by Dr. Paul the evidence of the existence of scrofula is of the most scanty, and often of the most doubtful character.

Lips.—The thick upper lip of the strumous which is never absent from the older descriptions of physiognomy is due, as before stated, to some irritation, usually to acrid discharge from the nose. A like cause will probably explain the thick alæ of the nose observed in some cases, and also a thickening of the inferior and anterior point of the nasal septum that is not uncommon in struma.

A red line on the gums has been described as an outward and visible sign of scrofula. Ruehle, describing the physiognomy of phthisical, speaks of its existence as among the features that would arouse a suspicion of phthisis. Various interpretations have been given to it, and Ruehle simply describes it as "a sharply defined red line at the edge of the gums, opposite the incisor and canine teeth."*

The teeth in scrofula show nothing distinctive. In many strumous individuals the teeth are normal in every respect, although, as occurs in other conditions of ill-health, dentition is often irregular or delayed. Often the teeth are uneven or projecting, or present milk-white spots indicative of defects in the enamel.† In other cases they appear to be brittle, are apt to scale off or break, and to become prematurely carious. These

---

* "Loc. cit." p. 510.
† D. Laycock, *Med. Times*, "loc. cit., p. 449.

various features are, however, by no means peculiar to scrofula, although often met with in that disease; and the condition upon which these modifications from the normal state depend are, I think, still unknown. One of the commonest aspects of the teeth in struma is that figured by Mr. Hutchinson,* who has given a drawing of the upper central incisors of the permanent set from "a very scrofulous boy." These teeth are very large, white, well-formed, and almost quite square. They are possessed, moreover, of a sharp, clean edge. Although not peculiar to scrofula, I have met with these teeth more often in that disorder than in any other.

Clubbed fingers are not common in the scrofulous. Trousseau had not met with an instance, although not a few authors have connected this deformity with the disease. Clubbed fingers, under whatever circumstances they occur, appear to be due to impeded circulation, to retardation in the return of venous blood, perhaps also to imperfect oxygenation of the blood. The condition appears to be most common in congenital heart disease,† in phthisis, empyema,‡ and chronic lung affection and certain thoracic aneurisms.§ In all these circumstances I imagine that the explanation just given would hold good. I have only met with one case of finger clubbing in scrofula, and that case was a very marked one. The patient was a boy, aged 10, with well-pronounced features of the so-called phlegmatic type of scrofula. There were a few enlarged glands on both sides of the neck, especially about its base, and some scars in the skin of less recent glandular mischief. The fingers of the left hand were very clubbed, those of the right being less conspicuously effected. In both axillæ enlarged gland could be felt that occupied the extreme apex of the space. These glands had not attracted the notice of either the patient or his mother, and formed

---

* "Illustrations o Clinical Surgery, vol. ii., 1879, plate 43.
† See figure in Dr. Laycock's article.
‡ Dr. Ogle. *Med. Times*, vol. 1. 1859, p. 291.
§ Dr. Meadows. *Med. Times*, same vol. p. 377. I quote these merely as examples.

masses of no great size. The collection on the left side was the larger, and formed a clump about the size of a duck's egg. The lad was the eldest child of five, and the only one of the family who presented traces of scrofula. The father died at the age of 42 of supposed phthisis. I would maintain that in this case the cause of the clubbed fingers was purely mechanical, and due to pressure exercised, most probably, upon the axillary vein, by the gland masses wedged in the apex of each axilla. The cold and purplish color of the hands quite supported this idea. Such a case as this appears to me to be parallel to one recorded by Dr. Ogle,* of St. George's Hospital, where a healthy man with an aneurism of the subclavian artery developed clubbed fingers on the affected side.

The General Manifestations of Scrofula.—Of the history of the scrofulous individual, of the tendencies to disease he exhibits, and of the actual affections to which he is prone, I can only speak in the briefest possible manner, and will attempt little more than an enumeration of the more common outcomes of the malady. Bazin has attempted to divide scrofula into four periods—primary, secondary, tertiary, quaternary; and has endeavored to establish a sequence in strumous manifestations akin to the very definite sequence of morbid processes observed in syphilis. Bazin's division, however, is arbitrary, artificial, and certainly without clinical foundation. His fourth period includes affections, such as phthisis, tubercular peritonitis, and amyloid degeneration of viscera, that cannot be classed as scrofulous without effecting a great and undesirable reform in the common significance of that term. As regards the sequence of diseases that his periods imply, it is possible that an individual may here and there be met with who would show such sequence; but at the same time I think it would be allowed that the circumstances were rather accidental than the outcome of a great pathological principle. Scrofulous disease may first show itself by some manifestation of Bazin's ter-

---

* *Med. Times and Gazette,* vol. i. 1859, p. 261.

tiary period—as, for example, by a joint affection—and then proceed to the secondary or primary disorders of his list, the patient becoming the subject of an eczema or a lupus. Commonly enough one finds a patient exhibiting throughout his whole life but one manifestation of one of Bazin's periods, and showing no other evidence of scrofulous disease. How often, for instance, does lupus form the sole manifestation of scrofula! and how often in strumous children is a carious spine the only outcome of the malady! Scrofula, moreover, often appears to be abortive; an infant presents an eczematous eruption abovt its head, or developes a phlyctœnular ophthalmia, and this may be followed perhaps by more or less gland enlargement, but there the affection ends—ends often abruptly, and the child grows up healthy and strong.

In opposing, therefore, in general terms Bazin's notion of the evolution of this disease, I would venture to assert that the progress of scrofula is most capricious and uncertain, its manifestations most variable, and untrammelled by any restriction as to order of appearing.

The various ailments of the scrofulous may, I think, be most conveniently dealt with according to the tissues or structure they involve.

Skin Affections.—It is doubtful if one can consider any skin disease, with the exception, perhaps, of common lupus, as peculiar to scrofula. The skin affections that occur in the scrofulous have, it is true, some peculiarities, and exhibit certain tendencies that may be more or less characteristic; but these few special features cannot place such affections in a special class, and cannot support any claim that they should be regarded as distinct forms of skin disease.

Those who, with Hardy* and Bazin,† maintain a peculiar form of cutaneous affection—the scrofulide—

---

* " Des differentes Formes des Scrofules cutanees ou Scrofulides." "Gaz. des Hop.," 1854, No. 115.

† " Art. Scrofulides, Dict. Encycl. des Sciences Med." Paris, 1880, p. 355.

must, I think, confess that the distinctions they lay down are somewhat scanty, a little artificial, and a good deal limited in their application.

One of the commonest skin disorders prone to affect the scrofulous is the *erythème pernio*, or *chilblain*. This affection particularly occurs in those patients who exhibit the so-called phlegmatic aspect of struma; and, indeed, in such individuals a history of chilblain is not commonly absent.

In many strumous patients there appears to be no difference between the chilblains they present and those that occur in non-scrofulous individuals. Commonly, however, the strumous chilblain assumes a more special aspect. It is apt to be associated with a greater amount of swelling, and with more thickening of the deeper parts than is usual in this form of erythema. It is apt, moreover, to persist, to relapse, and to lead to troublesome and prolonged ulceration. Chilblains will often continue to inflict strumous patients for nine or ten months out of the twelve, and I have seen cases where they have persisted all the year round. In such persons, also, the ulceration is often very marked, and may assume more the aspect of those ulcerating chilblains that are sometimes met with upon the toes of paralyzed limbs, especially in cases of infantile paralysis. I believe that erythema pernio of the fingers is much more common in the scrofulous than in individuals in any other condition of general ill-health.

Eczema is, next to chilblains, the commonest skin affection of the strumous. It very usually appears as the first manifestation of scrofula, and is of much more frequent occurrence between the first and second dentitions than at any other time. Its especial features—such as they are—are these:—It is apt to be chronic, and to show little tendency to spontaneous cure. It is seldom extensive, is little prone to recurrence, and selects as it favorite sites, the scalp, the furrow behind the ears, the concha itself, and the face, especially the parts about the lips and nostrils. In these latter situations it is often secondary to some discharge from the nose or some unhealthy condition of the mouth or tongue. The eruption itself is usually very moist, and its secretion tends to become pustular, leading thereby to that form of eczema known as eczema impetiginodes. When about the head or face it is as a rule associated with an enlargement of the cervical glands that seldom assumes any great magnitude. In strumous infants and young children this eczema is often followed by certain superficial abscesses that appear in or about the affected district as the eruption subsides. This especially applies to eczema of the scalp. These abscesses are very superficial, are covered by a thinned and purplish-red integument, and cause neither pain nor other inconvenience. If left alone they break and usually do well, or at any time their course may be very effectively cut short by a minute puncture with the thermo-cautery point.

These eczematous eruptions form a conspicuous feature in Bazin's "first period" of scrofula, and are prominent among the "benign scrofulides" of those who maintain a distinctive character for the skin disorders of the strumous.

Lupus.—It is impossible in this place to attempt even a sketch of this important skin disease and its various aspects. It constitutes one of the gravest outcomes of struma, and has received much attention as well from the pathologist as from the practical surgeon. I propose here to consider merely the question as to how far lupus is to be regarded as a scrofulous affection, as an affection peculiar to the strumous. The matter may be considered from both a clinical and a pathological point of view. Clinically it must be owned that the bulk of the cases of ordinary lupus occur in scrofulous subjects. I exclude, of course, from this question that eruption common in tertiary syphilis, and known as the "tubercular syphilide," or as syphilitic lupus. This disease certainly resembles common lupus both in aspect and to some extent in tendency, but that resemblance is only superficial. In syphilitic lupus we have —as Mr. Hutchinson would express it—an example of syphilis imitating a well-known skin-affection. It is an imitation merely, and upon the grounds of general pathology the two diseases are not identical. Exception must also be made in the present instance to the various forces of acquired lupus. This variety of lupus would appear to be due to accidental inoculation with certain animal matters in a state of decay, and occurs independently of struma or any other diathesis. The ordinary forms of lupus—untouched by these reservations—have great claims to be regarded as scrofulous diseases; and if there is a skin affection peculiar to struma I think that affection would be lupus.

Examples are certainly met with of lupus in individuals who present no familiar or recognized token of scrofula, but, as I have already pointed out, undoubted strumous affections may appear in persons who are of perfectly healthy aspect. A closer examination of apparent instances of lupus attacking non-scrofulous patients will give results such as these:—There is a history of phthisis in the family, or one or other of the parents was scrofulous in youth, or a brother or sister of the patient shows evidence of undoubted struma, or the patient himself has some of the tendencies of the scrofulous in a slight degree—a disposition to catarrh after trifling exciting causes, and a tendency for such affection to become chronic and relapse.

From a pathological point of view lupus shows a most close alliance to recognized scrofulous disorders. The lupoid process is chronic, is apt to relapse, extends locally, as if by direct infection of adjacent parts, and has products that show a marked disposition to degenerate. One important feature is absent—the tendency to gland implication; and the rarity of this complication is a great argument in the hands of those who deny the essentially scrofulous nature of lupus. Lupus, however, in spite of this, would appear to have—like other strumous disease—a predilection for lymphatic tissues; for it is very apt to occur about parts where skin and mucous membrane join, and to be very destructive in those situations; and it is well known that in such localities—as, for example, about the mouth and nostrils—the lymphatic networks are particularly numerous and extensive. Mr. Hutchinson,* moreover, has suggested that the lupoid process extends by means of the lymphatic channels of the part.

Microscopically, the changes in lupus are essentially scrofulous. In some cases distinct follicular tubercles are met with in the diseased area, and in other instances a less perfect formation of tubercle is the principal feature. In perhaps the bulk of cases, in the place of perfect tubercle a granulation tissue is met with associated with giant cells, and some few of the large cell elements so peculiar to scrofulous processes.† So far, indeed, as the microscope is concerned the lupoid process would appear to be actually identical with recognized strumous changes; and this fact, in addition to clinical and other evidence, would support the belief that lupus is as definite a scrofulous affection as scrofula itself is a definite morbid condition. This observation would apply only to ordinary lupus, and would not therefore imply that every form of lupoid disease is of an essentially scrofulous character.

The *Lichen Scrofulosorum* of Hebra, while it is not restricted to the strumous, is certainly most common among individuals of that class. In ninety per cent. of the cases of this skin affection observed by Hebra, either there were such distinct evidences of scrofula as gland disease, caries, and strumous ulceration, or the patients were the subjects of mesenteric disease, with its attendant mal-nutrition. This cutaneous affection occurs for the most part in young people between the ages of 10 and 25, and assumes the form of roundish groups of papules about the size of millet seeds, and of a reddish or faint brown color. Sometimes they are of the same tint as the adjacent skin, and according to Dr. Tilbury Fox may not occur in groups, but be more diffused in their arrangement.

These patches of eruption appear usually upon the trunk, tend to take on a very chronic course, and to frequently reappear. They are generally associated

---

* " British Med. Journ.," Vol. I., 1880.
† See paper by Dr. Grancher in *L'Union Medicale*. 1881, p. 874.

with certain spots, like those of common acne, and the whole eruption is, as a rule, most satisfactorily treated by general measures. Lichen and scrofulosorum appears to be of much greater frequency in male individuals, but is not limited to that sex, as was originally maintained. In an article in the "Transactions" of the Clinical Society of London, Dr. Tilbury Fox gives an excellent representation in chromo-lithography of this eruption, and furnishes details of several cases,* the bulk of which are oddly enough in females.

With regard to other skin eruptions, it has been frequently maintained that certain parasitic affections are of more common occurrence among scrofulous than among healthy individuals. This has been especially urged with regard to favus, and possibly the accusation is true. But I should imagine the relation between scrofula and parasitic skin disease to depend merely upon the circumstance that the conditions of squalor, dirt, and bad hygiene, so suitable for the development of scrofula, are also the very conditions that would most favor the growth and spread of parasitic affections. Favus, I believe, is almost if not quite unknown among the rich, and the patients among whom one meets with the disease are, as a rule, conspicuous for their miserable surroundings and their generally neglected condition. If a child has its head full of lice it is very reasonably assumed that its care has been neglected; and if the mother of such a child can so utterly ignore the practice of simple cleanliness as to allow its hair to swarm with vermin, she is probably no more careful about matters of diet, regimen, and ventilation—matters that hold a prominent position in the etiology of scrofula.

Skin affections, other than those already alluded to, do not appear to be more common or more peculiar in scrofulous individuals than they are in persons exempt from that diathesis, Lastly, certain rare cutaneous dis-

---

* Vol. xii, 1879, p. 140. See also case of Lichen S. in the same volume (p. 195), by Dr. H. Radcliffe Crocker. Hebra's account of the disease will be found in vol. ii. of his work "On Diseases of the Skin," New Syd. Soc., p. 52.

orders have been described as peculiar to scrofula that have in reality no connection with that disease, but are on the contrary manifestations of hereditary syphilis. As examples of these I may especially cite the so-called "impetigo rodens" and the "scrofulide rupiforme" of Hardy. Under these names have been described skin diseases that belong undoubtedly to infantile syphilis.

Scrofulous Gumma. Cold Abscess. Strumous Ulcer of Skin.—These variously named affections may be conveniently described together for reasons that will be immediately apparent. Sometimes in scrofulous subjects a little roundish indurated mass may be felt beneath the skin, and situate in the subcutaneous tissue. This minute hard mass when first noticed may be the size of a rice grain or a pea, or, on the other hand, as large as a cherry. At first moveable, it soon becomes adherent to the skin, and having slowly increased in size up to a variable point, it begins to soften in the centre, the skin over it becomes purplish, gradually thinned, and at last gives way. Through the opening so formed a thin curdy pus escapes. Such a small subcutaneous nodule when it attains a size no larger perhaps than a walnut is commonly known as a scrofulous gumma— especially by French writers.* Very frequently, however, the subcutaneous induration increases, the skin remains long intact and unchanged, and by subsequent breaking down of the hardened mass and extension of the process, a large collection of pus is formed; and then perhaps the term cold abscess would be used. The "*gomme scrofuleuse*" and the superficial cold abscess are indeed pathologically identical, and differ only in degree and extent. Microscopic examination shows that the subcutaneous mass is tubercular, that it gradually softens in the centre, having previously exhibited a caseous change, and by such softening forms an abscess cavity. The abscess enlarges by the invasion of the surrounding parts by its investing wall; and the microscope here also shows in this wall follicular tubercles

---

* For recent account see Art. by Brissaud and Josias, "Gommes Scrof.," *Revue mens. de Med. et de Chirurg.*, 1879.

and all the various grades of the tubercular process. Sir William Jenner* some years ago maintained the tubercular nature of these subcutaneous abscesses, and Lannelongue † has quite recently given a very perfect demonstration of the same view.

These superficial strumous abscesses may occur at any age. One of the most destructive I ever saw was in a man 25 years of age. As a rule they occur at a much earlier period, and are commonest during the first five or six years of life. Dr. Grancher states that they are more common in the acquired forms of scrofula—a statement I am disposed to endorse. They are often met with on the trunk than on the limbs, are common about the face and neck, and I think especially frequent upon the back. They may be single or multiple.

This tubercular process may be, of course, more deeply seated than the subcutaneous tissues, and is then very often associated with bone changes. The abscess formed under such conditions cannot be considered here, but it forms a collection of pus that in English text-books is more particularly designated "a cold abscess." Sometimes the indurated nodule is situated in the skin itself. In this position it never attains a large size, becomes soon apparent, and soon terminates by a discharge of the softened matter in its centre. To these purely cutaneous nodules the term " tuberculosis of the skin " has been especially applied, and Dr. Vidal‡ has elaborately described them, under the title of cutaneous " tuberculomes." This process is distinguishable from lupus by its more rapid course, its limited tendency to extend, and certain histological features.

It is from the bursting of a subcutaneous "gumma" that the scrofulous ulcer most usually results. A gland abscess may lead to such an ulcer, as may also a "tuberculome" of the skin, or even a pustular eruption on the skin, but the commonest cause is that first mentioned. In cases where these ulcers form, the integument has

---

* "On Tuberculosis." *Med. Times and Gazette*, vol. ii. 1861, p. 423.
† "Abces Froids et Tuberculose Osseuse." Paris, 1881.
‡ "L'Union Medicale," 1881, p. 639.

become much undermined before the pus has found an exit, and this condition of the skin, together with an actual extension of the tubercular process, is the chief local cause of the chronicity of these sores. The strumous ulcer needs no description. Its irregular outline, its purple, thinned and undermined edge, its base void of granulations, or crowded with those elevations rising up—large, pale, and flabby—is familiar enough.

A special form of cutaneous erysipelas has been described under the title "scrofulous erysipelas." It is said to occur independently of wound or injury, to conspicuously involve the lymphatics, to be more common in females, and about the age of puberty.* The details, however, furnished of this supposed special form of erysipelas are very scanty.

Mucous Membranes.—One of the commonest features in scrofula is a tendency to a catarrh, that may be induced by the most trifling causes, that is apt to persist long after the primary cause has ceased to act, and that is disposed to relapse and resist ordinary treatment. The mucous secretion in such catarrhs is generally thick and profuse, very prone to become mucopurulent, and often, it would appear, endued with irritating properties, as shown by eczemas, &c., of the skin about the orifices of affected mucous cavities. The more accessible mucous surfaces are, I think, attacked in this order of frequency—the pharynx, the conjunctiva, the auditory lining membrane, the nose, and the genitals. The mucous membrane of the alimentary canal and of the bronchi are also very commonly involved in the scrofulous. In the mouth and pharynx the most conspicuous feature is hypertrophy of the tonsils. This affection is almost pathognomonic of scrofula, for indeed in strumous children it is rare not to find some enlargement either of the tonsils themselves, or at least of the masses of adenoid tissue at the back of the pharynx. Such adenoid masses, it is well

---

* Grancher (loc. cit., Dict. Encyclop., p. 334) gives an account and references. See also account of case of tuberculo-caseous erysipelas, reported by M. Coigne, and detailed in *London Medical Record*, Feb. 19, 1873.

known, have a structure identical with that of the tonsil. These enlarged tonsils are nearly invariably associated with the appearance of a gland tumor on either side of the neck, and situate just about the top of the great cornu of the hyoid bone.

I presume that the almost constant presence of this gland, which feels rounded, deeply placed, and about the size of an hypertrophy tonsil, has given rise to the erroneous impression that enlarged tonsils can be felt externally.* So constant are these enlarged glands in this tonsillar affection, so uniform is their outline and situation, that it is almost possible to diagnose the throat affection by simply examining the neck with care.

Dr. West† states that hypertrophy of the tonsils commences usually during the later stages of the first dentition, and seldom attracts notice until the child is three years old. I believe that in many cases of marked struma it commences at an earlier date. I had under my care a male child, aged seven months, who was suffering from cervical gland disease, and who presented a very considerable enlargement of both tonsils. M. Robert‡ has observed the affection as early as the sixth month; and a systematic examination of the throats of strumous children will show that a distinct tonsillar hypertrophy is by no means uncommon before the age of two years. The "angina scrofulosa" of some French authors§ would appear to have little connection with struma, and to be in most cases either a syphilitic or a lupoid ulceration. Trélat‖ has given a good description of a tubercular ulcer of the mouth and tongue; but the disorder seems to be rather a feature of those patients who in the common clinical sense are classed as tubercular. The ophthalmic affections of the scrofulous

---

* For relations of tonsil to external parts, see fig. 11, "Bellamy's Surgical Anatomy," 1880, p. 41.
† "Diseases of Infancy and Childhood," 6th ed., 1874, p. 592.
‡ "Bull. General de Therap.," May, 1843.
§ Constantine Paul (*Gaz. hebdom.*, 1871, No. 47). Landrieux ("Arch. Gen. de Med., 1874, p. 660).
‖ *Note sur l'ulcère tuberc. de la bouche et de la langue* (Arch. Gen de Med., 1870, vol. xv., p. 35).

consist mainly in tinea tarsi and phlyctœnular ophthalmia. The latter affection is infinitely more common in the strumous than in any other class of individuals, and may almost be considered as peculiar to the scrofulous diathesis. Primary disorders in the deeper parts of the globe are not common in scrofula, and when they do occur in such patients appear to present no distinctive features. Otorrhœa is one of the commonest and the earliest manifestations of scrofula. The discharge is apt to become purulent, and is very commonly attended by some eczema of the auricle. Catarrhal otitis media is generally due to extension of mischief from the throat or nose. It is a serious affection, in so far as that it may at any time take on suppurative action, and may lead to necrosis of the petrous bone. I would venture to question the assertion of Birch-Hirschfeld * that "scrofulosis is at thebottom of the largest number of cases in which weakening or destruction of the function of hearing has taken place during the age of childhood." I believe, on the contrary, that the largest number of such cases would be claimed either by hereditary syphilis or be subsequent to scarlatina or other eruptive fevers.

Affections of the genital mucous membrane in young scrofulous females are by no means uncommon. The bulk of the cases of "infantile leucorrhœa," of blennorrhœa of the vagina, and of catarrhal vulvitis, occurring either in quite young children or about puberty, would appear to be associated with the scrofulous disposition.† Mr. Cooper Forster ‡ has described a "strumous ulcer of the vagina" as occurring in young children, but the evidence that such ulcers depend upon scrofulosis does not appear to be very convincing. The bronchial mucous membrane is commonly affected in scrofula. Strumous children often develop a troublesome cough on very trifling exposure, and may become the subjects of what has been definitely described as

---

* Loc. cit., "Ziemsen's Cyclopedia," p. 795.
† Dr. West, loc. cit., p, 756.
‡ "The Surgical Diseases of Children," London, 1860, p. 127.

"scrofulous bronchitis." Much attention has been paid to the pathology of this latter affection, as it has appeared to afford a connecting link between scrofula and phthisis. Such a link, however, is only demanded by those who consider the two disorders to be allied but not identical.* Affections of the alimentary tract show themselves by that much-discussed disorder—"strumous dyspepsia." That there is a form of mal-digestion at least common in scrofula, although perhaps not limited to that disease, must, I think, be allowed. This strumous dyspepsia particularly affects individuals of the "phlegmatic type," and shows itself by a capricious appetite, an almost constantly furred tongue, foetid breath, and a tendency to constipation, alternating with diarrhœa. In such persons the abdomen is often swollen, the skin pasty, the aspect dull and lethargic. In instances of this character one may reasonably assume that there is a catarrh of the alimentary tract not unlike that so common in more easily investigated parts, but there would appear to be little or no evidence to show that this catarrh is actually the cause, and in some cases the sole originator of scrofula, as has been urged by some.

Lupus may attack the mucous membranes, and may be either primary or have extended from the skin. It would appear to be of most common occurrence about the mucous lining of the palate, nose, and pharynx.

Gland disease will be the special subject of the second part of this book.

Diseases of bone form one of the most frequent and most serious of the manifestations of scrofula. The bone affection usually assumes the form of caries, and is characterised by the ease with which it may be induced, by its chronic course, its tendency to spread, to relapse, and to exhibit exacerbations. It seldom ends in resolution. This tendency to caries in struma contrasts markedly with the bone diseases of syphilis, where

---

* An excellent drawing, showing the changes in scrofulous bronchitis, will be found in Dr. T. H. Green's work on "The Pathology of Pulmonary Consumption." London, 1878, p. 27.

necrosis and periosteal affections are so common. The bulk of scrofulous bone diseases commence between the ages of 5 and 20. In the 509 Margate cases there are 152 instances of bone affection. These are distributed as follows; Bones of the foot, 35 ; the spine, 26 ; bones of the leg (mostly the tibia), 24 ; femur, 17 ; bones of the hand, 13 ; pelvic bones, 10 ; bones of forearm, 8 ; of skull, 8 ; humerus, 7 ; ribs, 2 ; and sternum, 2.

It must be remembered that these statistics apply only to in-patients, and the unduly low number assigned to disease of the bones of the hands is probably to be explained by this fact. There is one bone affection that would appear to be almost exclusively scrofulous. I allude to strumous dactylitis (the spina ventosa of the French). This disease usually invades the phalanges, and particularly the phalanges of the hand. It nearly always appears in young children under the age of five. Into the nature of the morbid change in dactylitis I cannot now enter, and will only observe that the disease appears to commence always in the interior of the bone, to gradually expand it, and ultimately to lead to its almost complete disorganisation. I have an opportunity of examining two cases microscopically, and in both instances the mischief had undoubtedly commenced in the interior of the bone. The flask-shaped appearance of the finger is most characteristic, especially when it is later on associated with a general reddening of the skin and the formation of several unwholesome-looking sinuses.*

Certain joint affections are common in the strumous, and are clinically known by the vague term "white swelling," or the more precise expression "fungous arthritis." The features of these chronic ill-conditioned joint diseases are well known. The morbid process is the same as that concerned in all other scrofulous manifestations, and some of the most typical examples of tubercle are to be obtained from the diseased syno-

---

* See "La dactylite strumeuse infantile," by Dr. Voquet. "Thèse de Paris," 1877.

vial membrane in these cases.* It would appear that in the majority of these joint affections the disease of the synovial membrane is secondary to changes in the articular ends of the bone.

Among the 509 Margate cases there are 143 instances of joint disease, which in order of frequency are thus localized:—Hip 65, knee 44, ankle 14, elbow 13, shoulder 3, wrist 3, sacro-iliac joint 1. This "order of frequency" exactly accords with that given by Birch-Hirschfeld and others.

The undue frequency of joint disease in individuals of the male sex has already been commented on.

The remaining affections to which the scrofulous individual is liable, include such as these—scrofulous orchitis, tubercular or strumous affections of mucous membranes, such as tubercular cystitis, "scrofulous tumors" of the brain, &c.

These latter diseases belong rather to the class known clinically as "tubercular," a term that must be considered as used somewhat in a conventional sense, and more as a matter of convenience than as indicative of any precise pathological meaning.

The term senile scrofula is applied to instances where familiar manifestations of the disease make their appearance in old people. Such instances are rare.

The ages of the subjects of senile struma have already been alluded to, and it would appear that the disease is more common in females than among men. In most instances there is a history of some scrofulous affection in early life, the main portion of the patient's existence having been, however, absolutely free from any trace of the disease. I might cite as an instance the case of an old man, aged 74, who came to the London Hospital with lupus erythematosus of the face, and extensive suppurating gland masses in the neck. He had been perfectly well up to the age of 70, but when a lad he had suffered from scrofulous ophthalmia, his cornea still exhibiting the opaci-

---

* Koster was, I believe, the first to describe tubercles in these joint affections. See Virchow's Archives, No. 48, p, 95.

ties of old ulceration. In this case the patient had been entirely exempt from any trace of strumous disease for a period of some sixty years. Often, however, the disease would appear to be primary, and such M. Bourdelais believes it to be in the majority of cases.* The diseases to which the subjects of senile struma are liable are identical with those common in the young, and among them may be mentioned ulcers of the skin, lupus, glandular disease, joint affections, and diseases of bone. M. Bourdelais places the last-mentioned diseases among the most frequent manifestations of scrofula in the old.

---

* "Sur quelques Observations de Scrofule chez le vieillard." Paris, These No. 297, 1876. See also " De quelques Lesions tardives de Scrof. chez les vieillards," by Dr. Du Moulin, Paris, 1854; and Sir James Paget's article already alluded to.

PART II.

SCROFULOUS AFFECTIONS OF THE EXTERNAL LYMPHATIC GLANDS.

"A SCROFULOUS GLAND."

As a preliminary step to this part of the subject it is needful to explain in general terms what is meant by a "scrofulous gland." It is obvious that such explanation must depend upon clinical features, and cannot, of necessity, be expressed in very precise terms. All chronically enlarged glands are not strumous, nor are all gland enlargements in delicate children of a certainty due to scrofula. This may seem the veriest truism and unworthy of mention were it not for the fact that many writers appear possessed of the belief that every weakly child with an enlarged gland in its neck must be the subject of strumous disease.

To assert that a gland is scrofulous these conditions should be present or at least frequent. The patient, probably a child, exhibits some of the general features ascribed to the scrofulous individual, or is actually suffering from some recognized strumous affection, or presents such a previous history, or such family tendencies as have been already described as usual in scrofula. The gland tumor will be most usually situated in the neck, will have increased slowly, will have been induced by some trifling peripheral lesion, and will have continued to enlarge long after that lesion has ceased to act. Moreover, the gland affection will spread in the absence of any fresh irritation, the mass will tend to become caseous, to break down, and to discharge its softened matter through the skin, leaving often peculiar sinuses or ulcers, and subsequently peculiar scars. The modifications of the process are infinite, but such are the principal features.

CHAPTER X.

AN OUTLINE OF THE ANATOMY OF THE EXTERNAL LYMPHATIC GLANDS.

It is needless to say that, apart from any scientific reason, it is essential to know the situation of the various external lymph glands, if only for diagnostic purposes.

Moreover, the bulk, perhaps all, of the gland affections of the strumous are secondary to some lesion or irritation at the periphery; and as the removal of this irritation is one of the first and most important elements in treatment, it is more than desirable to know from whence the radicles of the various glands are derived. I will therefore merely state, in the briefest psssible manner, first, the situation of the principal glands of the surface of the body; and secondly, the parts from whence their radicles or afferent vessels have their origin. This description is based almost entirely upon the admirable account of the lymphatic system given by Dr. John Curnow in his Gulstonian Lectures for 1879.*

HEAD AND NECK.

The following are the chief sets of glands:

1. The *suboccipital.*—One or two glands situate in the nape of the neck about the insertion of the complexus muscle. They receive the lymphatics from the posterior part of the scalp.

2. The *mastoid.*—Four or five small glands in the mastoid region. They receive the efferent vessels from the above set of glands, and thereby from a portion of the scalp.

3. The *parotid.*—Some five to ten glands placed some upon the surface and some deep in the substance of the

---

* *Lancet,* vol. i. 1879, p. 397 *et seq.* An exhaustive account of the lymphatics is given by Sappey in "Anat. phys. pathol. des Vaisseaux lymph." Paris 1871.

parotid gland. They receive lymphatics from the frontal and parietal regions of the scalp, from the orbit, the posterior part of the nasal fossæ, the upper jaw, and the posterior and upper part of the pharnyx.

4. The *submaxillary.*— Twelve to fifteen glands arranged along the base of the jaw under the cervical fascia. Receive lymph from the mouth, the lower lip and gums, and possibly also from other parts.

5. The *supra-hyoid.*—One or two glands placed in the median line of the neck between the chin and the hyoid bone. Receive lymph from the chin and median portion of the lower lip.

It is important to note that the efferent vessels from all these five groups of glands pass into the deep cervical lymphatic glands; a fact that considerably complicates any attempt to localize a peripheral lesion in many instances.

6. *Superficial cervical.*—Some five or more glands that are placed along the line of the external jugular vein, beneath the platysma, but superficial to the sterno-mastoid muscle. They receive, as afferent vessels, lymphatics from the auricle, from part of the scalp and skin of the face, and from the skin of the neck, and some of the efferent vessels of the mastoid and submaxillary groups of glands.

7. *Deep cervical, upper set.*—These glands, ten to sixteen in number, are placed about the bifurcation of the common carotid and along the internal jugular vein. They would be situated just above the upper border of the thyroid cartilage, and be also on a level with the hyoid bone. They receive lymphatics from part of the tongue, from the palate, larnyx, and lower part of pharnyx; from the tonsils, from part of the nasal fossæ, from the deep muscles of the head and neck, and from within the cranium. Efferent vessels from the parotid and submaxillary glands also enter this group.

8. *Deep cervical, lower set.*—These are situate in the supra-clavicular fossæ. They communicate with the axillary glands by a chain along the axillary artery and brachial plexus, also with the glands of the mediasti-

num, with those of the upper cervical set, and also with the sub-hyoid glands.

The cervical glands thus form a continuous series from the parotid and mastoid groups above to the subclavian and mediastinal below.

9. *Sub-hyoid.*—A few small glands are placed below the hyoid bone and above the middle line. It is said that the string of bronchial glands may extend up to this set. (Richet.)

10. *Retro-pharyngeal.*—M. Gillette (*Thèse de Paris*, 1867) described two small glands placed in front of the spine and upon the rectus capitis anticus major muscle. It is remarkable that into these glands certain lymphatics of the nasal fossæ enter. Hence, as Fraenkel * has pointed out, " retro-pharyngeal abscess may arise in consequence of disease of the nose."

It would perhaps be convenient to group the relations of certain glands to certain parts of the periphery according to regions.

Scalp.—Posterior part = sub-occipital and mastoid glands.

Frontal and parietal portions = parotid glands.

Vessels from the scalp also enter the superficial cervical set of glands.

Skin of face and neck = for the most part the superficial cervical glands. The lymphatics of the eyelids enter the parotid glands. (See observation by Sir William Jenner.†)

External ear = superficial cervical glands.

Lower lip = submaxillary and supra-hyoid glands.

Buccal cavity = submaxillary glands.

Gums of lower jaw = submaxillary glands.

Tongue.—Anterior portion = supra-hyoid glands and those near thyroid cartilage.

Posterior portion = deep cervical glands (upper set.)

Tonsils = deep cervical glands (upper set). Glands about angle of jaw and superior cornu of hyoid bone.

Palate = deep cervical glands (upper set).

---

* "Diseases of the Nose." Ziemssen's Cyclopædia, vol. iv. 1877, p. 187.
† *Med. Times and Gazette.* vol. ii. 1861, p. 423.

Pharynx.—Upper part = parotid glands and retro-pharyngeal glands. Lower part = deep cervical glands (upper set).

Larynx = deep cervical glands (upper set).

Nasal Fossæ = retro-pharyngeal glands, and some also beneath upper part of sterno-mastoid.* Lymphatics from the posterior part enter the parotid glands in part at least.

UPPER EXTREMITY.

The lymphatic glands of the upper extremity are—

1. *The Supra-Condyloid Gland.*—This gland is situated over the internal intermuscular septum of the arm, just above the inner condyle of the humerus. It receives some superficial lymphatics from the inner side of the forearm, and two or three fingers. Its efferent vessels pass up with the basilic vein to enter the lower axillary glands.

The Axillary Glands.—These glands are very numerous, and are arranged in two sets: (a) an internal set placed along the inner or thoracic wall of the axilla, and (b) an external set ranged along the axillary vessels on the outer aspect of the axillary space.

Some of the glands of the internal set are situated rather in the base of the axilla, being placed along the course of the long thoracic and subscapular vessels.

The axillary glands receive the superficial and deep lymphatics of the upper limb, lymphatics from the lumbar and dorsal regions, from the posterior part of the neck and shoulder, from the front and sides of the trunk, and from the mammary gland.

Some of the superficial lymphatics of the upper limb accompany the cephalic vein, and enter a gland just below the clavicle, or, passing over that bone, join the lower cervical glands. Sometimes two or three glands lie in the course of these vessels in the interval between the deltoid and pectoralis major (Curnow). If enlarged they may easily become the subject of a wrong diagnosis.

---

* For account of lymphatics of nose see Fraenkel's monograph, *loc. cit.*, p. 126.

LOWER EXTREMITY.

The following are the lymphatic glands of this part:—

1. *The Anterior Tibial Gland* (not constant).—This ganglion is situated in front of the upper part of the inter-osseous membrane. It receives such deep lymphatics of the leg as accompany the anterior tibial artery, and its efferent vessels enter the popliteal glands.

The Popliteal Glands.—Usually about four in number. One is placed superficially just beneath the fascia, at the point of entrance of the short saphenous vein. It receives lymphatic vessels that accompany that vein.

The remaining glands are deeply placed along the popliteal vessels. They receive such deep lymphatics of the leg as accompany the posterior tibial and peroneal arteries; and their efferent vessels pass up the limb with the femoral artery, and enter the deep set of the inguinal glands.

The Inguinal Glands.—These form a numerous cluster, and are divided into a superficial and a deep set.

(a) The superficial set average about ten glands in number, although Curnow has counted twenty in this situation. They are arranged in two clusters—one parallel and close to Poupart's ligament, the other parallel and close to the long saphenous vein. The former cluster, therefore, is almost horizontal the latter vertical.

(b) The deep set—about four in number—are placed along the femoral vein, and occupy the crural canal.

The inguinal glands receive the following lymphatic vessels:—

Superficial lymphatics of lower extremity. Enter the vertical set of superficial glands.

Superficial lymphatics of lower half of abdomen. Enter the "middle inguinal" glands of the superficial set (Curnow).

Superficial lymphatics of buttock.

Those from the outer surface of buttock enter the external glands of the superficial set.

Those from the inner surface of buttock enter the internal or vertical set of glands.

Superficial lymphatics of external genitals. Enter the horizontal set of superficial glands; some few going to the vertical set of the same glands.

Superficial lymphatics of *perinæum*. Enter the vertical set of superficial glands.

Deep lymphatics of lower extremity. Enter the deep inguinal glands.

The lymphatics that accompany the obturator, gluteal and sciatic arteries, and the deep lymph vessels of the penis, pass into the pelvis, and have no concern with the inguinal glands.

The efferent vessels from the inguinal glands enter the lumbar set of lymphatic ganglia.

Curnow states that in rare instances glands have been found in the following abnormal situations:—1. In front of the forearm. 2. Just above the umbilicus. 3. Over the seventh rib; and 4, along the inferior costa of the scapula.

CHAPTER XI.

THE ETIOLOGY OF SCROFULOUS LYMPHATIC GLANDS.

Exciting Causes.—All that has been said in the previous chapters as to the general etiology of scrofula applies of course equally to this particular manifestation of the disease. All that we are concerned with now is the especial etiology of gland disease, and the causes that actually incite or immediately induce that affection. Presume an individual to be a "scrofulous subject," that is to say, presume that he has either inherited or acquired a certain delicacy of health, a certain tissue weakness or defect, is it possible for such an individual to spontaneously develop grandular disease, or in other words, is this gland disorder of the strumous ever primary? The answer to this question must certainly be in the negative, with perhaps some very slight reservation. There is no doubt that in the great majority of all strumous gland affections the mischief is secondary, and is dependent upon some previous lesion of the periphery, from whence the lymph vessels going to those glands are derived. This point has now for several years been insisted on by most of those who have treated the subject in any way. In a recent communication to the International Medical Congress of 1881,* Dr. Clifford Allbutt, after asserting this very generally allowed fact, goes a step further and would maintain that cervical gland diseases " may be, and often are, set up in young persons by local causes *alone*." If local causes *alone* are sufficient to induce scrofulous disease, then scrofula may be induced in any perfectly healthy and robust person, if only the local irritation be properly directed. So far as one's present notions of scrofula are concerned this would appear to be a *reductio ad absurdum*. As the " local causes " to which Dr. Allbutt alludes are of a very simple nature and still more frequent occurrence ("irritation of the neighboring mucous membranes being the most common" of them), it would appear that an outbreak of gland disease may occurr in the neck of any individual, however vigorous; and it is remarkable that struma—if Dr. Allbutt's views be true—is not an almost universal disease. While it must be allowed that in the great mass of all cases some exciting cause is required to induce gland disease (*a tendency* to scrofulous disorders being already present), it must be confessed that in some few instances no local cause or peripheral lesion can be discovered. These cases, perhaps exist simply as an evidence of our want of knowledge ; and from what is known of the physiology and pathology of the lymphatic system, it is more than probable that in time the etiology of these few cases will be placed upon the same basis as the rest. The cases that exhibit this absence of apparent local exciting cause are very uniform in their chief characters. The

---

* " Abstracts of Communications," sec. iv. p. 106. "On the Origin and Cure of Scrofulous Neck."

patients usually present a distinct history of heredity, and especially a tendency to phthisis in the family. The gland disease increases very insidiously, and is usually somewhat wide-spread, being most common in the axilla and base of the neck; or in the groin also, in addition to those two situations. The gland masses do not tend to attain individually a great size, although collectively they form large tumors that often quite fill up one or both axillæ. If operated upon the glands shell out with remarkable ease, and exhibit the characters described in class II. (see below); that is to say, they show tubercular structure in great perfection. I have seen a good number of cases with features such as I have just described, and in none was there any suspicion of a primary local lesion. The only references I can find to this matter in books are these. Perrochaud* urges the existence of primary gland disease in scrofula, and states that such disease appears more often as a somewhat general adenopathy, affecting the neck, axilla, and groin. Birch-Hirschfeld† reports a case, almost without comment, of a lad aged four years who had a swelling of the axillary glands on one side as the sole evidence of scrofula. Not the least evidence of any peripheral lesion could be found. His father, it is important to note, had been a victim to scrofula in his youth. The axillary tumor suppurated, and then other manifestations of scrofula appeared.

I must repeat that these cases are most probably apparent rather than real exceptions to the general rule that scrofulous gland disease is never primary.

In searching for peripheral exciting lesions in any case it must be borne in mind that these sources of irritation are often of the most insignificant nature; and, moreover, the gland mischief may persist and increase long after all traces of the exciting cause have disappeared.

It is possible, for example, that the enlarged glands in the neck of a child of 10 or 12 might have been due to some defects in the first dentition, to some ulceration

---

* Quoted by Bourdelais, "loc cit.," p. 22.
† "Loc. cit.," p. 800.

of the pharynx that had occurred years ago, and had long since healed up. As an example of the persistence of gland mischief after the initial cause has ceased to act, I might cite this case from my out-patient department :—

The patient, a girl aged 13 years, had lost her father from phthisis, and had brothers and sisters who were afflicted with scrofula. The child was perfectly free from any trace of strumous disease until she was 8 years of age. She then accidentally ran a fork into her chin. The wound suppurated, and enlarged glands began to appear in the supra-hyoid region close to the puncture. In about three weeks the wound had perfectly healed, but the gland disease gradually increased and spread, until now she has large tumors in both sides of the neck, the glands in the supra-hyoid region having also suppurated and led to sinuses. There was a total absence of any other peripheral injury.

Here it would appear that gland disease was active five years after the lesion that first excited it had ceased to act.

Another point to remember in inquiries upon this subject—especially where the neck and axillæ are concerned—is that external gland disease may be set up by, or may extend from disease in the interior of the body. The cases of Drs. Hilton Fagge and Goodhart, already alluded to, illustrate this point, and show the extension of gland disease from the chest into the neck. In like manner, one knows that in phthisis the glands about the clavicle and also in the axilla may become enlarged. Through the kindness of my colleague Mr. McCarthy I had recently an opportunity of examining some gland masses that had been removed from the axilla of a woman suffering from phthisis. These gland tumors, which were very extensive, exhibited the precise characters of scrofulous glands. I have notes of two or three casses, where the gland mischief in the neck had evidently extended from like trouble in the mediastinum or about the bronchi, although, had the cervical disease not appeared, the gland affections in the latter situations would not have been suspected.

We might now consider the actual *local lesions themselves that induce gland disease.*

Enlargement of the *bronchial* glands is usually subsequent to bronchitis, especially the bronchial catarrh associated with measles. It may also follow upon whooping cough, and probably upon most lung affections. The *mesenteric* gland disease is directly caused by some catarrh (scrofulous catarrh) of the intestine, or by some graver lesion, such as ulceration of the mucous lining of the gut. It is seldom in post-mortems on scrofulous children over a certain age that one omits to find some enlargement of the mesenteric glands. The frequency of their occurrence would probably coincide with the frequency of those digestive disturbances that are so common in the strumous, although on the point of frequency few would be enclined to endorse the assertion of Wiseman that "whenever the outward glands do appear swelled you may safely conclude the mesenteric to be so too, they being usually the first part that is attacked by the malady." *

The peripheral disturbances that may induce gland disease in the *neck* are very numerous. Among them may be mentioned—eruptions of the head and face, especially porrigo of the scalp, ulcers of the skin, all forms of stomatitis, thrush, inflammatory affections of the gums or tongue, catarrh and ulceration of the pharynx, affections of the tonsils, coryza, ozœna, or other disorders of the nose, diseases of the external and internal ear, imperfect dentition, and decay of the first set of teeth. Of the comparative potency of some of these lesions of the surface I shall speak presently.

The great tendency for naso-pharyngeal catarrhs to produce gland enlargement can to a great extent explain the frequency of such enlargements after measles and scarlet fever. So constantly do we hear parents ascribing the scrofula in their children to an attack of measles that one would be disposed to doubt if a child with a scrofulous tendency could have measles and escape strumous gland disease, were there not instances to

---

* Richard Wiseman. "Chirurgical Treaties." London, 1692.

prove the fallacy of such a suggestion. In considering the disposition of measles to induce scrofulous manifestations, it must be borne in miud that this fever is one of the earliest ailments of children, that it furnishes perhaps the first opportunity that has been afforded of detecting any flaw in the child's state of health, and that it occurs, moreover, at a time when gland disease is naturally prone to appear. Gland tumors in the *axilla*, bend of *elbow*, *groin*, and *ham*, may be locally induced by various forms of peripheral mischief, such as eruptions, ulcers, broken chilblains, suppurative bone and joint affections, abscesses, etc.

So far the local affections that have been credited as causes of strumous adenopathy have been for the most part inflammatory; there remains to be considered the influence of injury and of cold, and the question of sudden gland enlargement.

Injury by inducing inflammatory changes may of course set up gland mischief, as in the case just reported, where a child was wounded in the chin with a fork; and examples of like nature are not uncommon. I have notes of a case, however, where I think the gland mischief was set up by an actual contusion of the gland itself. In this instance the child was struck across the side of the neck by a falling case, a swelling ensued, and on its subsidence an enlarged and tender gland was apparent. The disease subsequently spread in a leisurely manner. Price [*] asserts that sprains and other injuries may lead to primary enlargement of the axillary glands, but he adds nothing to the bare assertion.

Cold.—Dr. Reid observes, "I have seen the glands on one side of the neck and throat swelled and inflamed by a momentary blast of cold air." Other authors cite like cases, and many assign cold and exposure as direct exciting causes of strumous glands. I very much doubt if cold or exposure have ever directly induced a scrofulous gland. I can quite believe that such glands may follow upon exposure, but in such instances I imagine

---

[*] "On Scrofulous Diseases of the External Lymphatic Glands." By P. C. Price. London, 1861, p. 61.

that the lymphatic mischief is due to a catarrh of the throat or nose set up by cold or damp. It is well known how rapidly the cervical glands will often enlarge in a case of common sore throat.

I venture to doubt the reality of all the reported cases of so-called sudden enlargement. As an instance, I might quote the example furnished by Wiseman.* Says he:—

"I shall give you a remarkable instance of a cook's servant in the Old Bailey who, sleeping one summer night upon a form, his head slipping off the one side of this, his neck pressed upon the end of it. When he awakened his neck was full of struma on both sides, some as big as walnuts, others less. They were of different figures, and distinct one from another. He was frequently let blood and purged. All else was done that expert physicians and chirurgeons thought fit to relieve him, but the struma continued, and after a few days mattered, and became virulent ulcers. He died tabid within half a year."

One knows how long gland disease may remain dormant before it is perceived by the patient, and how often such discovery is accidental. In this case of Wiseman's and in like instances I should imagine that that the pain in the neck induced by an uncomfortable position drew the man's attention to the existence of the disease, which had been of previous standing. It is remarkable, in the present instance, that the "strumæ" appeared in both sides of the neck, and not only in the side injured or compressed.

Details as to age and sex—such as they are—have been given previously.

Situation of gland disease.—Taking into consideration the whole lymphatic system, the glands that are most frequently the seat of strumous disease are the cervical, the bronchial, and the mesenteric.

Considering only the external glands, an analysis of the Margate cases gives the following results. Out of a total of 509 cases of scrofula there were 155 ex-

---

* "Loc. cit.," No. 4. p. 400.

amples of gland enlargement. These were thus distributed:

Neck alone	131
Neck and axilla	12
Groin alone	6
Axilla alone	4
Neck and groin	1
Neck, groin, axilla	1
	155

These figures show the immense preponderancy of cervical adenopathy. They also show that the axillary glands are seldom affected unless those of the neck are also involved, and that gland disease in two remote regions (as, for example, the neck and groin) very rarely occur at one and the same time.

The nature of the peripheral lesion that induces gland disease.—So far as I am aware no satisfactory answer has as yet been given to the question—Why, of all other external glands, are those in the cervical region so often the seat of strumous disease? Many explanations have been offered. It has been said that the face and neck being uncovered are more exposed to such external influences as cold and damp. It has been said that the peripheral lesions that lead to scrofulous disease are more frequent about the head and neck than they are elsewhere. It has been said that the glands in the cervical region are more numerous than they are in the groin, axilla, ham, and bend of elbow. These explanations, I would hold, are not valid. Are not the limbs, the hands, and feet, subject to a vast number of inflammatory affections akin, at least, to those about the head and face? Ulcers are not uncommon upon the extremities of the scrofulous, the fingers and toes are often the seat of ulcerating chilblains, that remain inflamed for months, the feet and hands are by no means rarely exposed to wet and cold (if it be insisted that these agents have any direct action at all), and injuries, such as wounds, scratches, and abrasions are, I presume, more common about children's limbs than they are about their heads and faces. Even allowing that peripheral lesions are more common about the radicles of the neck glands than they are elsewhere, is their number so greatly in excess of those that can affect the axillary glands that the proportion can be expressed by the figures 131 to 4? I should imagine not. With regard also to the comparative numbers of glands in different parts, it might be acknowledged (no doubt correctly) that the cervical glands outnumber the axillary. But is this disproportion as 131 to 4? As a matter of fact some of the most extensive and most numerous gland masses met with have been turned out of the axilla, and there are many reasons for knowing that the glands in the neck are in no very great excess of those in the armpit. If frequent movement be considered to offer encouragement to the development of buboes in any part, then I think that the axilla and the groin would be less frequently at rest than is the neck. Lastly, I believe it will be acknowledged that the glands in the axillæ and inguinal regions are more often affected in the non-strumous than are the glands in the neck.

The explanation I would offer of this discrepancy is the following:—The peripheral lesions most active in exciting gland disease in scrofula are those that are located in the adenoid tissue of a mucous membrane. Adenoid or lymphoid tissue is to be found in the submucous tissue of most mucous membranes. But it varies greatly in amount in different parts. It is particularly collected about the mouth and pharynx, forming in the latter situation the largest masses of adenoid tissue in the body, viz., the tonsils. A vast quantity is also distributed over the posterior wall of the pharynx. It is plentiful in the bronchial mucous lining, and also in the intestines, where it appears as those special collections, the solitary glands and Peyer's patches, and has besides a very extensive distribution.

Now the most common sets of glands in the body to be the seats of scrofulosis are the cervical glands, the bronchial glands, the mesenteric glands; and it is at least significant that these organs correspond to the most extensive collections of adenoid tissue in the body, viz., those situated in the naso-pharyngeal mucous membrane, the lining of the bronchi, and the inner coat of the intestine. Entering more into detail, I should say that the commonest causes of cervical gland disease are situate in the pharynx, and especially in the tonsil. In non-strumous cases, one cannot fail to be struck with the extraordinary rapidity with which the glands are involved in inflammatory affections of the throat and tonsil. Indeed, one of the earliest evidences of such affections is often a tender and then a swollen condition of the glands about the hyoid bone. Compare these throat affections with an eruption on the scalp or an ulcer of the face—where no adenoid tissue exists—and note the tardy manner in which the glands are implicated. I have observed the glands unenlarged in instances where a sore upon the cheek or an eczema of the head has existed for weeks, and that, moreover, in a scrofulous child. One has sometimes an opportunity of comparing the effect of lesions in different sites upon the gland apparatus in the same patient. Such an instance is the following:—A little girl, aged six, a member of a strumous family, had a scrofulous ulcer upon the cheek that had existed for nearly a month, and yet the corresponding glands were not enlarged. She then became the subject of an ulcerative pharyngitis (*i, e.*, adenoid tissue became involved), and in a few days the glands on both sides of the neck were enlarged, and subsequently attained considerable proportions.

To make other comparisons. There is adenoid tissue in the nose, about the mouths, under the conjunctiva. There is none in the limbs. Note, then, how readily the neck glands become involved in strumous ophthalmia, in ozœna, in sores about the gums or tongue; and, on the other hand, observe how long a time may elapse before the axillary glands are implicated in a case of ulcer of the forearm, or in a case of dactylitis, or a case of suppurative mischief in a joint. How trifling very often is the gland enlargement in such instances. Moreover, this proness for adenoid tissue is quite in accord with what we know of the tendencies of scrofula, and especially of its disposition to involve lymphatic structures, and is quite in agreement with the definition proposed for the disease.

Putting aside this matter of lymphoid tissue there are some few other differences in the power of peripheral lesions to excite strumous adenopathy. It may be unnecessary to observe that a suppurative inflammation of the surface is more prone to involve glands than is a non-pustular affection; that an ulcer of the forearm is more potent than is a simple eczema in the same situation. These differences depend mainly of course upon the detail of time and the depth of tissue involved. Retained pus is an extremely potent cause of gland enlargement. For example, a porrigo of the scalp, where the scabs are allowed to remain and the pus to be pent up beneath them, is much more active than is a like disease kept free from scab formation. And this discrepancy has often appeared to me to exist independent of the extent of surface involved, the duration of the disease, and the depth to which it has extended.

Lastly, it must be remembered that individual idiosyncrasy has a great deal to do with the readiness or tardiness with which gland disease occurs after like provoking causes, and into this matter the bar of almost utter ignorance at present prevents us from entering.

CHAPTER XII.

THE PATHOLOGY OF SCROFULOUS LYMPHATIC GLANDS.

I propose to detail in this chapter the pathological changes that occur in strumous glands, and to give an account of the appearances they offer under the microscope.

The appearances presented by these diseased glands

vary greatly, not only when viewed with the naked eye, but also when subjected to a microscopic examination. But in spite of the many diversities in aspect that exist it must be borne in mind that the process that underlies them all is one and the same, and that throughout the whole series of scrofulous gland affections we have to deal only with the outcomes of that peculiar inflammation I have already described, an inflammation that numbers among its products elements so remarkable as giant cells and tubercles. It matters not whether the process concerned in this disease be called the scrofulous process or the tubercular process. so long as the fact is recognized that there is but one morbid change underlying the many aspects of the disorder, and that the apparent diversities indicate differences in degree rather than in kind.

I would venture to particularly insist upon this point in respectful opposition to the views of those who have maintained that several distinct pathological processes are concerned in strumous gland disease. By many these tumors have been divided into those that are due to simple chronic inflammation, those that depend upon a simple hypertrophy, and those that have been occasioned by a deposit of tubercle. Here three distinct morbid changes are credited with a capacity for producing the same diseased tissue—a strumous gland. If simple chronic inflammation can be the sole factor in scrofula, then the indolent bubo that may form in the groin of a healthy lad from an abrasion on the foot may be reckoned as scrofulous. With regard to hypertrophy, the conditions that are allowed to be active in producing hypertrophy elsewhere are, I think, not to found about strumous glands. The process would appear rather to be an irritative one; and if it be true that lymph glands accurately reproduce the features of disturbing lesions at the periphery, and those lesions be allowed to be inflammatory, it would seem that there is little need of invoking the aid of hypertrophy. Moreover, the description given of these hypertrophied bodies is that of a well-recognized variety of gland tumor that shows under the microscope changes very different from those of hypertrophy. I think, moreover that the glands described by Sir William Jenner* as being the seat of "albuminoid infiltration," do not merit the special character he ascribes to them. The account he gives of the aspect of such glands shows that they are absolutely identical, in aspect at least, with an indolent gland enlargement in struma that shows the microscopic changes incident to scrofula, and affords no evidence of any peculiar infiltration.

In detailing the minute changes observed in this affection, I propose to divide scrofulous gland tumors into two classes. There is no object for this division other than that of a greater convenience for description. The two classes represent but grades or degrees of one and the same morbid process. In actual fact they are separated by no rigid line of demarcation, but merge imperceptibly the one into the other, and the division I have placed between them is therefore purely arbitrary. In the first division will be described the changes occurring in those glands that exhibit the process in its more active or intense forms, and in the second division those enlargements that show the more indolent change, and that are in a sense the more chronic.

† The first change noticed in scrofulous glands is an increase in the number of their lymph-corpuscles, an increase observed both in the sinuses of the organ, and in the proper gland-tissue itself. The more active the process the better marked is the change, the sinuses in such cases appearing so blocked and crowded with lymphoid bodies, that Berlin blue injected by a Pravaz's syringe only imperfectly follows the course of the lymph-paths.

It is significant to observe, that an increase in the number of their contained corpuscles is the very first indication of the inflammatory process as it effects lymphatic vessels, a condition constantly observed by

---

*Article on "Rickets." *Medical Times and Gazette*, vol. i. 1861, p. 260.
† The account that follows of the histology of these glands is in the main a reprint of that given in a lecture I delivered at the Royal College of Surgeons in March 1881, and that was published shortly after in the "British Medical Journal."

Dr. Klein in the pleural lymphatics of animals, the subjects of chronic pleurisy. And in connection with this point it is necessary to remember that the mode of development of lymphatic gland masses shows that they are merely modified lymphatic vessels. Whence these corpuscles come it is perhaps at present impossible to say, but it has been suggested, with reference to this condition both in the glands and in the vessels, that they may have been derived from some initial seat of inflammation, or that they are leucocytes escaped from the neighboring blood-vessels, or finally, that they owe their origin to the proliferation of the existing cells of the part. And it may here be convenient to allude to a singular and incorrect statement that appears to have been handed down from one pathological handbook to another; and that is, to the effect that the changes in the glands are first observed at their extreme periphery, *i.e.* in the cortical sinuses and follicles, a statement explained by the theory that the periphery is that part of the gland that would be first exposed to any irritant brought from a primary seat of inflammation. With reference to this statement, I must assert that in no single instance, in all the specimens examined, have I ever found the morbid changes commencing at the periphery of the organ, including in that term the cortical sinuses, and a fair amount of the cortical follicles. Indeed, the exact converse holds goods, and the changes invariably commence in the medullary or deeper portions of the gland. It must have been obvious, over and over again, to all who have cut open a scrofulous gland, that the gross changes visible to the eye are always most advanced in the central parts of the organ. If there be only a few specks of caseous change visible, they will always be found to occupy, not of necessity the precise centre of the affected organ, but, at least, a part other than the extreme periphery. The same holds good for limited suppuration, and, within certain limits, for even extensive purulent or caseous collections.

I have a section cut from a large gland of more than one inch in diameter that was absolutely caseous throughout, with the sole exception of the extreme periphery, where unaltered lymph-corpuscles can still be seen in great numbers. If, however, the previous statement were true, and the changes commenced in the periphery, one would expect to find them most advanced in that situation. So much for the fact. With regard to the theory, notice has not apparently been taken of this fact, that the afferent vessels of the gland first enter the plexus between the strata of the capsule, and then pass on to the cortical sinuses. The capsular plexus, therefore, is the part of the gland first brought into contact with infecting material, and yet the capsule is acknowledged by these very observers to be only secondarily affected; aad, as a matter of fact, this plexus is almost the last part of the true gland-tissue to be involved.

The next change noticed in these glands assumes the form of certain spots of varying size and shape, and which owe their conspicuousness to the fact that, in specimens prepared with hæmatoxylin, they appear more lightly stained than does the rest of the gland. They are observed first in the medulla, then in the cortex, and are always limited to the gland tissue proper. The more active the process the more diffused is the change, and the appearance of distinct spots or foci of altered tissue is best seen in glands that do not exhibit the most vigorous aspect of the disease.

Closer examination shows that these districts are occupied by a large number of cellular forms of very diverse character; they range from ordinary lymph-corpuscles on the one hand, to certain very large and conspicuous cell elements on the other. These latter are nucleus-like roundish bodies, that stain lightly, are pellucid, and present a very distinct intranuclear plexus; between them and the unaltered lymph-corpuscles every gradation can be observed, and there can be no doubt that from these latter they are derived. All these elements, both large and small, show evidence of active division. The endothelial cells of the part appear unaltered. The blood-capillaries in these districts are very numerous and distinct, and are often provided with a well-marked adenoid sheath that, while described by His as of normal occurrence, is certainly best seen in slightly inflamed glands. That these conspicuous changes indicate so many spots or foci of inflammation may, I presume, be granted, especially on the ground of their precise resemblance to changes elsewhere of undoubted inflammatory origin.

The large cells with glistening protoplasm form a great feature in all scrofulous inflammation. They were first noticed by Rindfleisch, who regarded them as characteristic; he described their segmentation, and conceived their origin to be from the leucocytes of the part, an origin almost undoubted in the present instance. All subsequent pathologists have observed them, and their almost uniform appearance in scrofulous inflammation is an important factor in that process. With regard to the future of the cell-products in these exudations, it can only be said that, as a whole, they are short lived, or become involved in subsequent changes. The large elements, especially, soon undergo degeneration and disappear from the scene; in no instance do they experience any further development, and they have no share whatsoever in the formation of giant-cells, which, it may be remarked, are never observed at this stage of the process. The changes indicated become more or less general, and very soon fresh manifestations of a somewhat different character become apparent.

These concern the lymph sinuses of the part engaging first those situate towards the medulla. The endothelial cells that line these passages show evidence both of individual and of the numerical increase, and become very distinct. At the same time the fine reticulum that normally occupies the lumen of the sinus becomes considerably augmented by the development of fresh fibres and bands that stretch across the passage, and help to more fully occupy it. The fibres of this new reticulum are soft and less fine than is the normal tissue, and some of the bands that stretch across the sinus are rather membrane-like expansions. This reticulum is closely associated with the lining endothelial cells, and from them it is probably developed.

This change is very similar to that observed by Klein in the lymphatic vessels under the pleura, and in the lungs in cases of chronic pleuritis. Within the lymphatics in this inflammation he describes a reticulum as being developed, and to which he has given the name of the "endo-lymphangeal network." *

Within these altered sinuses are a number of leucocytes, some normal of aspect, others exhibiting the changes already described, and a small proportion of the large cell elements of Rindfleisch. Other tubes or passages become apparent in glands the seat of strumous dsease, that I have ventured to describe as lymphatic vessels proper to the gland tissue.† They also develop across their lumen a fine reticulum, and show changes identical with those exhibited by the sinuses.

Marked changes at the same time are going on in the gland-tissue, adjacent to these affected vessels and sinuses. That tissue, already crowded with altered lymph-corpuscles and the typical large-cell elements, becomes conspicuous by the appearance of opacity, and the loss of the transparent condition of the intercellular spaces. This opacity is always first noticed in the immediate vicinity of the sinuses or vessels, and is due to deposit in the part of a homogeneous material that is obviously only coagulated lymph. The corpuscles appear embedded in this material, and the anatomical details of the affected district become indistinct. The fibres of the adenoid reticulum increase in density and width, become softened, and of so nearly the same refractive index as the structureless coagulum, that the general ground-substance of the spot assumes a more or less homogeneous appearance. These spots tend to entend by peripheral increase, and always assume, therefore, a rounded outline; for the same reason, a faintly concentric arrangement is often given to such relicts of fibrillation as persist. The trabeculæ become altered, their whole tissue becomes infiltrated with lymphoid corpuscles, their fibrillation very indistinct, until at last

---

* The Anatomy of the Lymphatic system—the lung. London, 1875.
† For account of these vessels, see *British Med. Journ.*, loc. cit.

they cannot be distinguished from the adjacent gland-structure, especially as by this time every trace of the corresponding sinus will have been lost. In this manner spots of opaque altered tissue appear simultaneously in many parts of the gland; they increase in size still retaining their rounded outline, often fuse together, and so produce those conspicuous patches that have been so prominently regarded by all who have investigated this subject. The cell-elements in these patches soon indicate a condition of decay. The largest cells are the first to perish, and are soon lost in the general monotony of fatty degeneration; whereas the endothelial cells are the last to go, and, even when caseous changes are far advanced, will persist, shrivelled and deformed, as the sole survivors of a once vigorous crowd of corpuscles.

This caseous change will be more fully described subsequently, but the process does not usually end here; on the other hand, the caseous districts soften and break down, and suppuration follows in the manner to be immediately detailed.

Such are the changes that occur in glands that exhibit the scrofulous process in its most intense or least chronic form. It will be observed that the process runs its course, and terminates in caseation without either giant-cells or tubercles having made their appearance, and yet it will be allowed that the changes noted are peculiar and distinctive.

When the scrofulous process is more indolent or less intense, the appearances as shown under the microscope are somewhat different to those just described; and this difference depends to a great extent upon the fact that the more active or intense the inflammatory process the greater is the tendency for the products of that inflammation to the cellular rather than fibrous. So far cell-elements have been most conspicuous in the process; but in the somewhat less vigorous forms now to be described, it will be noted that a development of fibrous material becomes a marked feature.

The process is similar, differing only in degree. The earlier changes assume more distinctly the character of well-isolated spots than they do in the form of the disease just cited. The reticulum within the sinuses is denser and more extensive, as is also the adenoid tissue of the proper gland-substance. The result is, that the opaque basis or ground-substance of the patches is now less homogeneous, and shows more distinct fibrillation; the change being slower, the cellular elements show more varieties of outline, and, on account of the prominence assumed by the fibrillar structure, appear to be much less numerous. Now it is in the midst of such opaque patches that giant-cells first appear. The aspect of these bodies is well-known. It is necessary to insist that they never, as some have stated, commence the process; they indicate the attainment of a certain stage in the inflammation, and never appear, under any circumstances, until that stage has been reached.

These rounded opaque patches with their giant-cells form one of the most conspicuous features in scrofulous gland disease, and have been made very familiar to all pathologists by Cornil, who has given to them the title of "îlôts strumeaux."

One word here as to the nature of these giant-cells. I would urge that the so-called giant-cells are merely lymph-coagula involving in their coagulation a greater or less number of cellular elements. The reasons for this belief are the following: the material of which the mass of the so-called cell is constructed is precisely similar to undoubted lymph-coagula. These bodies are found, under favorable circumstances, distinctly to occupy the lymphatic sinuses of the gland. Their position in these channels is not clearly indicated in the bulk of instances, owing to the rapidity with which the anatomical details of the part are lost. They precisely resemble in every respect the giant-cells sometimes found in chronically inflamed connnective tissue, and that are located without doubt in the lumen of lymphatic vessels. Their advent is associated with the appearance, throughout the greater part of the gland, of a material precisely similar to coagulated lymph, and that could hardly, from its extent and distribution, be regarded as protoplasm.

In opposition to the theory that they are protoplasmic masses, I would remark that the number of contained nuclei bears no relation whatsoever to the size of the cells. The arrangement of the nuclei in some of these bodies is so casual and incoherent as to suggest no other explanation than that of a plug of cells in a coagulum. Then, again, if the cell-elements in the neighboring tissue are few, the nuclei of the giant-cell are few, and *vice versâ*. Giant-cells are found in a tissue on the point of death and decay, and in parts that are absolutely, if not quite, non-vascular; and I cannot conceive that, surrounded by these conditions, it is possible for such active and vigorous changes to ensue, as must of necessity exist, if the protoplasmic theory be correct. It is urged by some that they are developed in blood-vessels; but, as just stated, they do not appear until the tissue has become practically non-vascular. The orderly arrangement sometimes noticed in the nuclei clustered at the margin of a giant-cell is considered conclusive as to the origin from the endothelium of a blood-vessel; but a similar orderly arrangement can be seen in giant-cells occupying lymphatic vessels, whose endothelium is not disturbed.

I believe that these masses are formed in the meshes of the reticulum that occupies the sinuses, and also in the gland tissue proper. In the latter situation they may be located in the lymphatic vessels, to which I have briefly alluded. They indicate the absolute cessation of all lymphatic current, for, up to the time of their appearance, the possibility of a circulation of lymph still exists. It will be remembered that the first changes appear in the medulla; its sinuses are blocked, the efferent vessels appear quite empty, and yet the cortical sinuses are distinct, often dilated. This appears to support the belief that lymph is still entering the gland by the afferent tracks, and may be actually distending the organ, owing to the greatly obstructed outflow. I believe that their processes are merely portions of the reticulum in which they are deposited; and it is a most significant fact that the distinctness and density of the processes vary precisely with the state of the surrounding tissues. These processes are acknowledged to be continuous with the adjacent network; and so similar are the refractive indices of the reticulum and the cell-mass, that that part of the reticulum embedded in the mass is not obvious. The shape, moreover, sometimes assumed by these bodies is eminently suggestive of a formation within a vessel.

In advancing this view as to the nature of giant-cells I am alluding only to the giant-cells of scrofulous or tuberculous processes, and I would maintain it for the giant-cells incident to all forms of these processes.

In the "Lancet" (vol. i. 1879) will be found an excellent summary of the various views upon the subject of giant-cells. From this account it will be seen that many pathologists incline to the opinion that these masses are formed within vessels, and I would point out that the size of these bodies militates against the idea that the are formed within blood capillaries. Klebs conceives their origin to be from coagulated albuminous bodies in the lymphatics. Köster and Hering maintain their origin from the endothelium of the same vessels, and Lübinow* suggests that some at least are formed within lymph tubes. Lanceraux,† speaking of tuberculosis of lymphatic vessels, states that the lumen of the vessel becomes occupied with cell products that form the centre of the tubercle, and that the passage of lymph is thereby arrested by a veritable lymph thrombosis.

To return to the changes in the gland. As the process advances caseation begins in these opaque patches of altered tissue or "ilôts strumeaux." This change is preceded by certain alterations in the cell-elements of the part that has been described by Grancher as vitreous degeneration.‡ It is not, however, very well marked in the caseous process in scrofula, but is best seen—according to Grancher—in caseous pneumonia.

---

* Virchow's "Archiv. Band.," lxxv., Heft. i.
† "Traité d'anatomie pathologique," vol. ii. part i. 1879, p. 487.
‡ "Dict. Encycl.," loc. cit. p. 307.

The affected cells become greatly swollen as if by some colloid (vitreous) change. Their protoplasm from being granular becomes homogeneous and clear, the nuclei waste and are soon lost, and the cells themselves in time fuse together into a compact mass. One effect of this change is to give to the part a somewhat gelatinous aspect. Caseation follows. The caseous process is but a form of fatty degeneration accompanied by some desiccation of the part. It commences in the centre of the affected tissue, and proceeds towards the periphery. The leucocytes and their products disappear early, the giant-cells resist the change for some time, and the last cells to be lost are the endothelial cell-plates. The fibrous tissue of the part survives for a while the simpler elements, and is then lost in the uniformity of the change. All that is to be seen in a caseous spot are granular *débris* and fatty matters, with here and there perhaps some shrivelled relics of what was once a cell, and some faint fragments of a fibrous material. Caseation is generally associated with some development of fibrous tissue in the adjacent parts, with a species of sclerosis, and the more indolent the cheesy change, the more distinct is this development. It is by the formation of this fibrous matter that caseous masses may become in time encapsuled, and rendered to some extent inert. Later still the caseous districts begin to soften in the centre. This change is probably more or less a chemical one, and has been compared to the softening of old cheese. By the extension of this liquefying process cavities are formed filled with a creamy kind of matter. In some cases this may probably dry up, but more often suppuration is excited in the adjacent parts, and the collection becomes purulent. Thus arise the bulk of glandular abscesses. On the other hand, it must be noted that the caseous process may pass on to a calcareous change, and the whole gland become converted into a mortar-like mass.

The naked eye appearances presented by glands of this class vary considerably. In the earliest stages the gland when cut open shows a pale flesh-colored section, and is soft and uniform in density. As the disease advances the color becomes somewhat paler, and the texture of the mass of greater firmness. The tissue, moreover, has a clear, semi-transparent aspect, so that when a small gland mass is held up to the light it appears fairly translucent, especially about its periphery. Moreover, when caseous nodules exist they can often be detected at some little depth from the surface of the section, and if the gland tumor that contains a cheesy nodule be small—no larger, for example, than a horse bean—the little mass can often be recognized by its opacity when the uninjured gland is held up to the light. Before cheesy nodules are, however, apparent, some parts of the gland become of a paler color; and loosing their translucency look dull and more opaque than the tissues around. This alteration may be limited to very minute specks scattered over the surface of the section, but more often it is restricted to one or two patches of fair size. These larger patches are generally indistinct in their outlines. The color of these parts becomes paler and paler, their aspect more and more opaque, until they assume the appearance of caseous change. The caseous districts vary greatly in the features they present. Usually, they are extremely well limited, and project sharply above the cut surface of the gland when it has been sliced open. They vary in color from a dull white to a faint grey or a fainter yellow, and often the central part of the spot is of a different color to the periphery. The size of the caseous spots varies. They may be as small as hempseeds or large enough to occupy almost the entire gland. They may be single or multiple, and exhibit the greatest irregularity in outline. There is no connection between the size of the gland and the extent of the caseous change. A little gland no larger than a horse bean may be wholly caseous, while a tumor one inch in length may show on section but a few minute specks of that degeneration. In more advanced periods of the disease purulent cavities appear in the gland, which when fully formed are filled with a creamy kind of pus of a peculiarly greenish color, very like that of a duck's egg. Some of the more chronic of these glands may attain great size before they caseate, and I have no doubt that it is to these homogeneous tumors that the term hypertrophy has been applied, especially as they retain for a long while more or less the aspect of the normal gland.

2. The glands placed in this division are those that show the more indolent phases of the scrofulous process. The whole morbid action in these bodies is leisurely, and serves to exhibit the most perfect attainment of the strumous change, of which indeed they exhibit the least intense aspect. Chronicity is hardly the term to use in comparing various changes in scrofulous disease. The question of time is one too uncertain to be of use as a means of comparison, as the gland disease is in all cases apt to progress with the utmost irregularity.

In the glands placed in this division a production of fibrous tissue is the most conspicuous element. In the other forms of strumous gland just described there is a tendency for the morbid tissues to assume more or less rounded outlines, and to arrange themselves in the form of spots or patches. So in the present instance it will be found that the fibrous matter so conspicuous in these bodies is prone to display itself in more or less rounded masses, and to produce in consequence some very definite appearances.

In some cases, these rounded masses are so dense as to appear quite solid, the contained cell-elements being scanty and withered; they can be isolated by shaking, and, although they present no characters that would now distinguish them as tubercle, yet in less recent descriptions they have received that name. In other districts will be seen a rounded spot, occupied by a more or less open but irregular fibrous structure, that often presents a concentric arrangement at its edge. Such an area contains different degenerate cell-elements, the periphery and adjacent tissue being at the same time probably occupied by lymph-corpuscles, but little altered. Throughout all parts of the affected spot can be seen the homogeneous material that closely resembles coagulated lymph. In other instances, giant-cells are introduced into these spaces, and the appearance denominated tubercle is produced. Into the reputed structure of the true reticular, lymphoid, or submiliary tubercle, I have already entered. It is sufficient to refer to the fibrous material so arranged as to include a more or less circular space, with a giant-cell in or about its centre possessing branched processes that reach the periphery: and casually to mention the larger and smaller cell-masses that occupy the tissue between the giant-cell and the periphery, and that are considered to represent all gradations between that cell and the little lymph-corpuscles that may crowd the outskirts of the tubercle.

Many of the affected districts, however, contain no giant-cells; others only show the circular arrangement of the fibres, or show giant-cells with no definite fibrous arrangement; while another part contains few, if any, of the corpuscles that are considered typical of the appearance. Certain is is that, with the appearance of this tubercle, the process ends, and degeneration of a gross character immediately ensues.

I would urge for the giant-cells found in tubercle a like nature to that I have ascribed to those bodies in dealing with the previous class of gland tumor. The circular outline often noted in these tubercles appears to be merely a circumstance in the inflammatory process, by no means peculiar or specific. The anatomical details of the sinuses are no longer obvious in gland-tissue presenting tnbercle; but I would still urge that these giant-cells are lymph-coagula formed in the irregular meshes of the now quite disordered stricture. Some districts show no giant-cells, but in their place the material that is in no way different from a coagulum. The masses, moreover, are as often seen at the periphery as in the centre of the supposed tubercles, and the smaller giant-cells would represent smaller coagula. The connection of the processes of one giant-cell with those of another, and the still further connection of those processes with the surrounding reticulum, is fully explained on the supposition that these giant-cells are coagula deposited haphazard in the meshes of the fibrous tissue of the gland, and that they conceal the reticulum to an extent equivalent to the size of the coagulum or so-called cell.

In proof of this I would call attention to the figure of a giant-cell taken from the edge of a caseous patch This giant-cell and the surrounding tissues are degenerating; and when such degeneration takes place, the cellular elements and the lymph-coagulum are always the first to perish, the fibrous elements resisting for a longer period the caseous action, and acquiring thereby a temporary distinctness. Therefore, if the giant-cell is deposited in a reticulum and bolts out by its very substance the details of that reticulum, it is obvious that, as the mass degenerates, the fibres of that mesh-work should again become apparent. Now, in this decaying giant-cell, a fibrous mesh-work can be seen stretching across the mass, exactly similar to that observed in the vicinity; and especially it is to be noticed that this reticulnm is continuous with the so-called processes of the giant-mass, and, through them with the adenoid tissue in the neighborhood. The appearance of fibrous matter in moribund giant-cells has been observed. Klein, in speaking of the giant-cells in tubercle of the lung, states that, before undergoing final decay they are often converted into a fibrillar substance; on this point, however, I would urge that the development of fibrous material in a tissue partly decayed and quite destitute of blood-supply would seem scarcely probable. Slowly a caseous degeneration spreads over these glands, and in more advanced instances that ends in suppuration.

The naked eye appearances presented by these bodies on section are at first somewhat akin to those described in the former examples of the disease, although they never appear so vascular nor so brightly colored. In time the section becomes dull and opaque, and an insidious cheesy metamorphosis creeps over the diseased tissue. The suppuration when it occurs is more diffused, and throughout there is always evidence of abundant fibrous material in the parts.

This fibrous matter may be so extensive as to render the gland tough and firm in some portions of it, and to cause it even to creak under the knife when cut.

In all cases of gland disease in struma, there is more or less thickening of the capsule, but in the present instance that thickening reaches its greatest development. As will be seen in the chapter that deals with local treatment, this increased density of of the capsule is of good service when an attempt is made to remove the diseased contents of the body by scooping.

Those who, like Cornil, maintain a distinction between the scrofulous and tubercular processes, as they term them, would designate the gland affections dealt with in Class I. "scrofulous," and those dealt with in Class II. "tuberculous." This distinction is to a great extent a mere matter of terms, inasmuch as there is great unanimity in the accounts given of the morbid appearances by those who hold the most opposite views as to their nature. An examination of many glands in various stages from the same case will show that there is no line of demarcation between the changes incident to glands of the first class and those of the second. The glands that show tubercle present in their earlier stages appearances identical with those seen in glands that show Cornil's "ilôts strumeaux," although in the former instance those appearances are much modified by the greater indolence of the process.

It has been asserted that in the so-called "scrofulous gland" the change commences in the connective tissues of the part, and is a true interstitial adenitis, whereas in the "tubercular gland" it commences in the lymphatic vessels and sinuses, and is a species of catarrh.

Such distinctions, I must say, are not obvious. Take a case of gland disease that would accord with Cornil's description of the "scrofulous gland," and among the masses removed will very possibly be found a few minute glandular bodies that are just beginning to be invaded by disease. I have examined many such, but never have I observed at the commencement of the process any changes in the connective tissue elements.

The clinical distinctions between these two classes of

gland tumor are generally well marked. The glands of the first class enlarge tolerably quickly, although the matter of time must not be too strictly weighed, as at any stage the process may remain quiescent for a long period. Their progress may be marked by inflammatory symptoms. They are apt to attain large size, forming, indeed, the largest individual gland tumors of scrofula. They tend to become matted together. They caseate early, and show a very fairly constant disposition to suppurate.

Those of the second class increase in a very indolent and insidious manner; their progress is marked by an utter absence of any inflammatory symptoms. They may form by extension large gland collections, but the individual glands are seldom of great size. There is no periadenitis, but the tumors remain freely moveable. Caseation appears slowly, and these glands show but a very slight disposition to suppurate, and that only in advanced cases. The cases not infrequently met with of persistent and well-marked enlargement of one gland (or at the most of one or two glands) belong, I think, in every instance to the first division of gland enlargements.

One matter concerned in the pathology of this affection remains to be considered. The manner in which strumous gland disease spreads. Some cases of spreading gland disease may be explained by assuming that the peripheral irritation that caused the first gland to enlarge is still active, and is spreading along the other lymphatics to other glands. Other cases may be explained on the grounds that many glands may be simultaneously affected from one source of irritation. But these explanations will not cover all instances of extensive or extending gland disease in scrofula.

One constantly meets with cases where glands are involved, one after the other, in a very regular succession, and in a direction often actually the reverse to that of the lymph-current. From some definite peripheral irritation a gland may enlarge just above the clavicle, and be followed, in the absence of any fresh irritation, by a series of glands that will appear in order one after the other, and mount up the neck, the last gland affected being that nearest the jaw. In such a case, it may be urged that the circulation of lymph is so interfered with by the implication of the orignal gland, that a species of backward stasis is produced, and the diseased lymph thrown thereby into collateral channels. In the same way it is very, common to see gland-mischief extend from the neck into the axilla, in a most precise order; and by direct continuity of parts.

The microscope, however, reveals another explanation; the diseased process may actually be followed from the gland-tissue back into the capsular plexus, and thence into the afferent vessels: these latter become blocked up by the corpuscular elements, develop within their lumen a reticulum, just as observed by Klein * in inflamed lymphatics elsewhere, undergo all the scrofulous changes, and become in time caseous. I imagine that, by means of these vessels, connected glands may be affected, especially as one finds connecting lines of diseased tissue running from one gland to another. Moreover, this morbid change can extend along the vessels, independently of valves, and in the reverse direction to the lymph current.

Then, again, it is well known that the larger glandular masses may present more separate gland-bodies than can be accounted for by the anatomist. Many of these are no doubt due to the enlargement of glands so small in health as to escape the eye of the dissector. And, again, I have repeatedly found little glands wholly caseous, which are yet no larger than hemp-seeds. Now, I believe that many of these apparently supernumerary bodies are produced from the afferent lymphatics of the adjacent diseased glands by the following process. Klein has shown that irritated lymphatic vessels very readily develop into masses of adenoid tissue surrounded by a perfect sinus, which sinus is nothing indeed but the altered vessel itself. These little masses may assume the features of a gland

---

* "Anatomy of the Lymphatic System," loc. cit.

(although they retain a much more rudimentary structure), are quite capable of increasing indefinitely, of showing the whole of the changes incident to scrofula; and I have no doubt that they form, in time, many of the bodies that are not to be distinguished from the ordinary lymph-glands.

This tendency to local infection or to spreading of disease by direct continuity of tissue is of considerable importance in discussing the operative measures proposed for the treatment of these gland affections.

CHAPTER XIII.

SYMPTOMS AND DIAGNOSIS OF SCROFULOUS LYMPHATIC GLANDS.

Scrofulous gland tumors exhibit the greatest variety, not only in their physical character, but also in their progress and tendencies.

In one individual the most diverse conditions of gland tumor may be observed. The same patient may exhibit masses more or less advanced in destruction by the side of minute glands that show but the earliest evidences of disease, and may be the subject at one and the same time of gland enlargements that have progressed with rapidity, and of other enlargements that have been practically quiescent for years.

The commencement of the gland mischief in scrofula is usually very insidious. Most commonly the enlarged bodies are discovered by accident on the patient casually passing his hand over the neck or other affected part. Sometimes the gland tumor reaches considerable size before it is discovered, and this is especially the case with disease of the axillary glands. This casual discovery of masses that have already attained large size is probably the foundation of many of the accounts of reputed sudden glandular enlargement. In nearly all cases there is at first an utter absence of any pain or tenderness in the part, and of any of the ordinary signs of inflammation, the process being essentially indolent and chronic. In some rare instances the gland disease may commence with symptoms of active inflammation, a circumstance observed usually in quite young children and in cases of the disease that have followed upon the eruptive fevers. In the latter instance the activity of the inflammation depends probably rather on the exanthem than upon any scrofulous influence. As a general rule it may be said that the more marked the evidences of struma in any patient the more disposed will the gland affection be towards a chronic course; and I think it is true that the examples of more active gland disease are for the most part met with iu those who exhibit but slight proofs of the scrofulous disposition. It may, however, be here observed that in extensive and spreading gland disease the invasion of a new cluster of glands is occasionally marked by some pain and tenderness and some undue heat in the part.

Before dealing with the grosser forms of gland disease, it is well to note that in making a careful examination of the neck in some strumous children who appear free from lymphatic affection, one may often detect (especially along the posterior triangle) a number of small, distinct, hard, and freely moveable glands that should not be obvious in health. These slightly enlarged glands may persist for some of the earlier years of the child's life, and while they are apt to temporarily enlarge during any disturbance of the patient's health, they soon return to their normal size again; and, giving no further trouble, disappear at puberty.

Strumous disease may appear simultaneously in many glands, or may commence in one and remain for a long time limited to it. I have notes of cases where only one gland has been affected, which gland has attained fair size, and undergone the whole scrofulous change without the appearance of any fresh disease in the vicinity. The affected glands in any case are recognized at first as small, distinct, and very freely move-

SCROFULOUS LYMPHATIC GLANDS. 147

able bodies, that readily slip under the finger. They are round or oval, but most commonly bean-shaped, and feel very firm and resisting, and often remarkably hard. Certain of these glands increase in size, and attain the dimensions of a filbert, or even of a bantam's egg, and yet remain perfectly mobile; and, although softer than when first noticed, are nevertheless elastic, firm, and throughout of equal density.

The number of such glands may increase considerably, and extensive chains of diseased masses be formed, the common characters of the tumors being still retained. Thus in the neck large, irregularly lobulated tumors may be formed that are moveable, painless, covered with healthy skin, and that can be felt to be made up of a number of distinct and but feebly adherent lymphatic bodies. The morbid changes in such tumors will be probably such as are described in the chapter on their pathology under Class II. The course of this variety of the glandular tumor is usually very indolent, and the enlargement commonly proceeds in an irregular manner.

Other glands increase somewhat more rapidly, and not infrequently with some local evidences of inflammation. They may form very large tumors, and produce considerable deformity. Neighboring glands become matted together, and so great lobulated masses are produced, that become adherent to the deeper parts, and subsequently to the skin. These glands soon present to the touch different degrees of density, and become softened in places. In time, in most instances, suppuration will ensue, of which more will be said subsequently. These are the glands that become caseous comparatively early, and it is important to note the fact of their becoming adherent is evidence of their containing some purulent or quasi-purulent collections. The adhesions formed by these tumors are due, as already stated, to an inflammation set up in the soft parts about the gland. If only single glands are enlarged they usually follow the course just detailed, and as regards actual numbers, fewer glands are involved in these masses than is the case with the species of gland disease previously mentioned.

Speaking generally, the gland affections in scrofula are not symmetrical, and if symmetry be observed in a few instances it is probably but a coincidence. An exception, however, is afforded to this rule in certain cases of gland enlargement depending upon hypertrophy of the tonsil. Here symmetrically placed glands are often to be felt in the neck at the level of the hyoid bone. Glands that simultaneously commenced to be enlarged do not usually all progress at the same rate and in the same manner. In a chain of diseased glands that were at one time all of the same size and condition, some will in a while be found much enlarged and advanced in disease, while others are still but little altered. And in most instances it is impossible to detect any cause for this irregular progress. As may be supposed, the glands that are the nearest to any source of peripheral irritation are the ones most severely affected, but even this relation would not appear to hold good in every case.

The progress of strumous gland disease is, as already noted, most variable and uncertain. Glands may enlarge rapidly and then become quiescent, and remain stationary for indefinite periods, that may sometimes be estimated by years. Or, on the other hand, enlargements that have pursued an indolent course from the commencement may abruptly take on more active change, and speedily end in suppuration. Glands that have attained certain dimensions may subside more or less, and then enlarge again; and this phenomenon may be exhibited more than once in the same set of gland tumors. There is no doubt that this local affection is considerably influenced by the general health of the patient. At puberty a marked improvement is often observed in gland affections that have given much trouble in childhood, and this improvement not uncommonly amounts to perfect cure. In adults the state of the local mischief is greatly influenced by the patient's condition, and this is especially observed in connection with pregnancy and parturition. These conditions are often attended by the outbreak of a gland disorder that had perhaps been long quiescent.

In some cases an absolute reappearance of gland mischief will occur in connection with defects in general health. Such cases may be illustrated by this example. The patient, when a child, had scrofulous glands in the neck that suppurated and left conspicuous scars, but no trace of any remaining glandular enlargement. He was free from any trace of struma until the age of twenty-two, when large gland tumors again appeared in the neck, and led to extensive suppuration in less than twelve months. Previous to the onset of this fresh disease, the patient had been very dissipated, and had in consequence suffered greatly in health.

Resolution may occur at various stages and in many different kinds of gland. Those cases where children exhibit a slight but uniform and very chronic enlargement of many glands (the largest, perhaps, not larger than a hazel nut) usually do well, and at puberty all trace of the disease may disappear. Single glands more commonly undergo resolution than do the collections made up of several glandular tumors. It is difficult to say how far strumous disease may advance and still cure without suppuration be possible. It is certain that gland tumors of large size that have become matted together, and that are known to contain at least quasi-purulent collections, may subside and shrink, and cease forever to give the patient trouble, although there may be always some evidence of their presence. Certain of such glands become calcareous and thereby inert; but that final change is less frequent among the lymphatic bodies of the surface than it is among those within the cavities of the body. I think that the mediastinal and bronchial glands furnish the largest number of instances of calcareous change.

Resolution is least common in those insidious glandular tumors that increase within certain limits, and yet spread along a whole chain of lymphatics, and that, on microscopic examination, are found to contain well-developed tubercle. Resolution, or cure without suppuration, is much more common in children than in adults; and, indeed, in the latter it is very rare.

Cases are recorded of gland enlargements having disappeared under the influence of measles, scarlet fever, and angina,* and a like happy result is stated to have occurred after an attack of erysipelas of the face.†

Suppuration.—The majority of scrofulous gland tumors at some time or another end in suppuration. What percentage thus terminate it is impossible to say. Reliable statistics upon this point can scarcely be attainable, inasmuch as cases are not long enough under observation; and, moreover, few surgeons can enjoy a practice that would include a sufficiently large number, both of the most trifling and of the most serious examples of the disease.

Those who have written upon the subject have come to the most diverse conclusions. Price‡ states that suppuration occurred in 82 cases out of 140 examples of cervical gland disease that came under his notice, and observes that it would probably occur in time in the remaining cases. On the other hand, Phillips§ remarks that " of twenty persons suffering from sensibly enlarged glands, in scarcely more than one will they proceed to suppuration." In the 131 examples of cervical gland disease obtained from the Margate records, suppuration had occurred in 93 instances.

The following are the principal facts about suppuration in scrofulous lymphatic affections. Suppuration is infinitely more common in the external glands than in those in the interior of the body. It is also more common in the superficial series of external glands than in the deep. The greater number of instances of inert caseous glands and of calcareous change will, so far as the neck is concerned, be found in the most deeply seated of the cervical lymphatic bodies. From what has been already said it will be gathered that suppura-

---

* Des adenopathies chez les scrofuleux. " These de Paris, 1877," by Dr. Legendre.
† Two cases are given by Deligny, "loc. cit.," p. 66.
‡ Loc. cit., p. 72.
§ Loc. cit., p. 11.

tion occurs with much greater frequency in adults than in children. Suppuration is, however, often early and severe in gland disease in infants. With regard to the period in the disease at which suppuration occurs, it is impossible to speak precisely. I think the average period may be estimated by years rather than by months. Some slight light is thrown upon this matter by noting the duration of non-suppurating gland affectious. The Margate cases show that the average duration* of such affections is 3.5 years; the maximum period being 12 years and the minimum a matter of months. It is certain that suppuration occurs as a rule at an earlier period of the disease in adults than it does in children. I observe, also, that healing occurs more quickly when the glands have suppurated early than when they suppurate late, a fact that may be well surmised. Observations I have made in a large number of cases establish the interesting fact that the presence of other strumous manifestations evry markedly delays suppuration in the glands.

Local Evidences of suppuration.—There are two distinct forms of "glaudular abscess" in scrofulous cases. In one instance the suppuration is in the gland itself, and is limited by its capsule up to a certain point; in the other example, the suppuration is in the connective tissue outside the gland (peri-adenitis), and has of necessity no communication with any purulent collection within the tumor. It is important when possible to distinguish between these two abscesses. To take a typical example of each form.

1. *The gland abscess proper.*—A gland mass that has existed for some time and has attained good size has become adherent to the deeper parts and to the skin. From being hard it becomes of unequal density, and feels less resisting in places. The outline of this gland is distinct, although before the pus escapes it may be obscured by an œdema set up in the adjacent parts. The mass becomes tender and painful, and the part hot.

---

* By this term I mean the leng'h of time the disease had existed when the patient came under notice.

The skin over the most prominent portion of the tumor is red and œdematous. It then becomes more and more thinned and of a purplish tint, and at last gives way and allows the pus to escape. Before the skin has yielded a sense of fluctuation may be evident. It is, however, of limited extent, and gives rather the impression of a soft and elastic spot in the midst of the denser substance of the gland mass. The pus is usually thin, and contains curdy grumous fragments that crush under the finger, and that represent undissolved portions of the caseous mass. If the opening can be examined, at the bottom will be seen the ragged interior of a disorganized gland tumor, and perhaps much unaltered cheesy matter. The discharge continues so long as any diseased tissue has to come away; then granulations arise, and in favorable cases the abscess cavity is filled up and the sinus closed,

2. *The abscess outside the gland.*—This generally occurs in connection with processes of some activity, and the abscess may form around a gland of still small size and not yet adherent to the skin.

Owing to the inflammatory changes in the soft parts around it, the outline of the gland is soon entirely lost, and the ordinary evidence of an abscess in the subcutaneous tissues become apparent. These evidences are often obscured by the presence of other gland tumors in the vicinity. The part becomes tender, and hot. The skin becomes red over a larger area than is affected in the former instance, and when it gives way often does so by a larger opening. Fluctuation is much more evident than in the case of the abscess within the gland. It can be detected over the whole area of the tumor, and is not bounded by an indefinite area of harder and more resisting tissue. The pus is laudable and of normal aspect, and contains no cheesy fragments. If the aperture be enlarged, a diseased gland will be seen exposed at the bottom of the abscess cavity. This cavity as a rule will not close until the gland has been more or less entirely destroyed by natural or artificial means.

In actual practice the distinctions between these two

kinds of abscess are often much obscured. Sometimes the two forms are combined in one case, and often the gland abscess proper is associated with much inflammatory change in the parts around, even although that change may not lapse into actual suppuration.

Not uncommonly the skin sloughs over the suppurating gland, and this obstruction may be extensive. A very usual complication is the undermining of the skin before the pus has found an exit. This leads to troublesome sinuses and intractable ulcers. The undermined skin about the oriffice of the abscess cavity is thinned, purplish, and of poor vitality. Where it joins healthier parts a tubercular process will often extend in the subcutaneous tissues and add daily to the mischief. This mode of extension is identical with that observed in the walls of a cold abscess. In nearly all cases there is a tendency for the opening to become fistulous if the pus has been allowed to find an exit for itself, and very often there are several of such fistulous apertures. These sinuses and these ulcers with undermined edges are apt to become very intractable, and give trouble for an indefinite period. Some of the worst cases I have seen have been in adults. Sometimes these glandular abscesses reach the surface with a remarkable abscence of inflammatory symptoms, and may often be classed with the most frigid of cold abscesses. Very often when one gland has suppurated and been thus eradicated, another will come forward and repeat the process, which may thus be prolonged, almost indefinitely. These gland cases afford many examples of what Sir James Paget has termed "residual abscess," that is to say, an abscess occurring from the remains or residues of a previous supperative process. A gland suppurates, and before, perhaps, all the disease has been eliminated, the process ends, the materials dry up, and the sinus closes. The case may appear one of cure, and for months or years the part may seem entirely sound. Then, probably from some defect in health, an abscess will appear in the old spot, and there is every reason to believe that it has arisen from the residues of the previous trouble. Often, too, I believe purulent collections within glands dry up and remain quiet for indefinite periods, then a residual abscess forms, and the matter is discharged through the skin.

Scars.—The cicatrices left after the healing of suppurating glands and the closure of old sinuses and ulcers are often very conspicuous. In the neck especially are these scars apt to produce much disfigurement. and to form a permanent evidence of the scrofulous disposition in a person, They vary greatly in appearance. In some cases a number of nipple-like processes or minute pedicles of skin are attached to the scar. In other cases, the cicatrix is marked by bars or ridges of altered skin that are unpleasantly conspicuous. These ridges are generally covered with a very thin, shining, purplish integument, like purple tissue paper. This delicate covering is usual in scrofulous scars, and especially in those of recent date. Other scars are indented and corrugated, or show evidence of much contraction, and thus resemble the cicatrix of a small but deep burn. Very often the altered tissue is firmly adherent to the deeper parts, and the cicatrix becomes thereby depressed; and in some cases this depression is very conspicuous. The color of the scar tissue is usually different from that of the surrounding parts. This difference is marked in recent cases, and, although it becomes less pronounced in time, may always be obvious. The color is that of a dusky red or purple. On exposure to cold the purple tint becomes greatly exaggerated, and the least excitement or unwonted exercise will intensify the red blush in the scar. In warm weather these scars will freqnently appear very conspicuously red. In many cases they are very sensitive and tender, and remain so for years, the condition varying often according to the state of the patient's health. Sometimes a part of the scar will give way and some discharge of pus take place, or it may become the seat of ulceration, or the thin cuticle having given way it may exude a thin serous discharge. These complications as a rule are coincident with some defect in the general condition of the patient. As years pass on the scar may become somewhat less conspicuous, chiefly by its losing some of its unnatural color, and assuming more the tint of the adjacent skin. Its more prominent parts, its bars and ridges, may atrophy, and so another source of disfigurement be removed ; it may also become less adherent than it was in the first instance, and drag less upon the parts around. Traces of it will, however, persist until the last days of the patient's life.

Pressure effects.—Instances of injurious compression upon neighboring structures are more often afforded by the mediastinal glands than by any others. One of the commonest ill-results of such compression is a perforation of the trachea, as a rule about its bifurcation.* Examples of injurious pressure effects are not often afforded by the external glands, and such examples as there are have for the most occurred in the cervical region. Among the most frequent structures to be pressed upon are the jugular veins. In such cases the face may look bloated and purplish as if from extreme cold, and this aspect, combined with the large neck full of gland tumors, gives the patient a very "apoplectic appearance," if one might be allowed to accept the conventional idea of the aspect of those disposed to immediate apoplexy. Dr. Deligny states that cerebral hyperæmia may be produced by this pressure on the veins.† I have notes of several cases of severe epistaxsis in connection with cervical gland disease. In one or two instances, the relation between the bleeding and the gland tumors is obvious, and I imagine that the connection between the two depends upon pressure on the venous trunks. I might cite the following example that occurred in one of my out-patients at the London Hospital :—

A girl, aged 16, had a considerable number of enlarged glands in both sides of her neck. She presented the so-called phlegmatic type of struma. These

---

* See case recorded by Dr. H. Thompson. Clinical Lectures. London, 1880, p. 39. And another by Mr. Edwardes. *Med. Chir. Trans.*, vol. xxxvii. 1854, p. 151.
† " Loc. cit.," p. 78.

glands were first observed two years ago, and appeared after an attack of scarlet fever. They had already suppurated, and when I saw her there were old scars on both sides, and three open sinuses on the right side of the neck. For the last twelve months she had been troubled with epistaxis. Her nose would often bleed three times a week, and then perhaps there would be no hæmorrhage for two or three months. The epistaxis was always severe when the gland tumors were at their largest, *i. e.* just before they suppurated. As soon as they had broken, the patient was for a while free from any attacks of hæmorrhage. She had menstruated regularly.

Dr. Deligny asserts that the carotid vessels may be so compressed as to produce cerebral anæmia or even ulceration of the wall of the vessel. He also gives references to cases of injurious pressure upon the sympathetic nerve trunk and upon the vagus nerve. Examples of pressure upon the former nerve are not uncommon. In not a few instances the recurrent laryngeal nerve has been so compressed as to have its function interfered with. Several cases have been noted of pressure upon the trachea and larynx producing alarming results. Mr. Cooper Forster * gives a drawing of a case of immense gland disease in the neck of a child aged five. The patient died of suffocation. David Craigie † gives a case from Bleuland of an infant whose deglutition was impeded by the pressure of enlarged glands upon the gullet.

Leucocythæmia is very rare in strumous gland cases. I believe that it only occurs in instances of rapid enlargement of many glands, and may be present to some trifling extent in the first stage of other cases. Although the glands are crammed with leucocytes these do not pass in any numbers into the general circulation. This fact can be established by a microscopic examination of the efferent vessels of the gland as they

---

* Surgical Diseases of Children, p. 101.
† Elements of General and Pathological Anatomy, 1848, 2nd ed. p. 287.

leave that body at the hilus. In the earliest stages of the disease sections of these vessels show certainly that they contain many lymph corpuscles, and probably in some cases a greater number than in health. But in more advanced cases of the disease, especially when the stage of caseation has been reached, the efferent vessels are found almost empty or presenting but a very few leucocytes among the coagulum that may partially occupy their lumen. This obstruction to the passage of lymph may be explained by the early occlusion of the medullary sinuses. (See Chapter XII.)

In Chapter XII. will also be found an occount of the manner in which gland affections spread.

Dr. Craigie, in the work just alluded to, describes what he terms a "strumous mortified bubo." This consists in a sudden enlargement of the "glands at the bend of the arm." The skin over them soon gives way by sloughing, and a deep, foul sore forms, with sharp cut, irregular edges. At the bottom of this hole is the diseased gland. After the destruction of the gland and much sloughing the part heals. Cruikshank observes, "I have known the last-mentioned glands (the brachial) die and slough out in scrofula without any great inconvenience."* A "syphilitic-strumous bubo" is described by Fournier as a scrofulous degeneration of a gland already affected with syphilis. He says it is not uncommon in cases where syphilis has attacked a strumous person, and is most often observed in the groin, although it may appear in any part.†

Diagnosis.—From the account already given of these gland tumors little has to be added under the heading of Diagnosis. Strumous disease in glands may be mistaken for simple and syphilitic bubo, and for the tumors that characterize Hodgkin's disease (the malignant lymphoma of Billroth, and the lymphosarcoma of Virchow). Among the other affections that have been named in connection with the differential diagnosis of strumous glands are various solid and cystic tumors, and the manifestations of glanders and farcy.

---
* Anatomy of the Absorbing Vessels. London, 1790.
† Nouveau Dict. de Med. et Chir. Prat. Art. " Bubon."

The chief points in the diagnosis of the scrofulous tumor are these:—The age of the patient (most often in children), the site of the mass (most often in the neck), the history of the patient, the existence possibly of other strumous disorders, or the evidences of past outcomes of the disease, the indolence of the affection, its trifling exciting cause, its persistence, and its tendency to caseation and the formation of pus. The diagnosis between scrofula and the earliest stages of Hodgkin's disease is often difficult, often for a while impossible. The main elements in differentiating between the two affections are these. In Hodgkin's disease the enlargement is rapid, the affection spreads with marked persistence, several sets of glands in various parts of the body may be simultaneously attacked, and there is an absence of periadenitis, of cheesy degeneration and of ready suppuration. In Hodgkin's disease, moreover, there is soon to be noted anæmia, emaciation, muscular debility, and a general and rapid failing in health. It would be impossible within the limits of this chapter to attempt to lay down the differential diagnosis between scrofulous gland tumors and various other growths, solid and cystic. Such diagnoses depend upon general principles, and open up too wide an area in surgery to be entertained in this place. As a matter of fact, if Hodgkin's disease be excluded, the diagnosis of a scrofulous gland is in ninety-nine cases out of a hundred a very simple object to be attained.

CHAPTER XIV.

THE TREATMENT OF SCROFULOUS LYMPHATIC GLANDS.

General Measures.—It is needless to observe that before any local treatment is adopted for the relief of scrofulous gland disease general measures for cure must be made use of, and means applied for the improvement of the general health of the patient. Scrofula is more than a merely local affection, it implies a serious deviation from the normal state, and expresses itself by certain tissue defects that are by no means limited to any one part of the organism. It must be confessed that these general measures, which involve some of the most elementary factors of a state of health, are often seriously neglected, and not infrequently give place to some favorite local plan of treatment.

It is to a great extent, if not entirely, useless to prescribe medicines and advise applications for a case of gland disease if the patient on leaving the prescriber returns to some squalid habitation, where he will be surrounded with the very conditions that have caused and maintained his disease. In many instances, the general hygienic measures needful for the proper relief of scrofulous affections cannot be fully carried out, but that is no reason why in other cases those measures should be neglected, and the onus of a cure thrown upon purely local modes of treatment. It is the merest truism to say that nothing in the past has contributed more to the lessening of strumous diseases than has the improvement that has taken place in the hygienic surroundings of the poor; and no treatment of scrofula will be well founded unless it places in the first position a regard for the general health and circumstances of the patient.

The first indication in the treatment of scrofula is simply this, to surround the patient with the best possible hygienic conditions. These conditions would comprise plenty of fresh air and light, good ventilation, a generous and properly regulated diet, suitable clothing, exercise in the open air, and a judicious culture of the skin.

I am aware that these conditions are not within the reach of a vast number of the scrofulous poor, but for some at least they are obtainable in various charitable institutions.

And I think the manner in which some of these valuable institutions are made use of is a strong criticism upon the value attached by many to general hygienic measures for the relief of scrofula. What kind of cases are to be found in these excellent charities? Cases of incipient disease that, commencing in some city slum, can be cut short by sea air, good food, and a healthy dwelling? Such are the cases that should be found, but they are very rarely to be met with. On the contrary, the bulk of the cases in these institutions are examples of advanced disease, cases that are convalescent from a grave malady rather than cases where that malady is being warded off. If the importance of good hygiene in the treatment of scrofula were only more fully recognized, then would these charities be available for the prevention of disease rather than for the patching up of advanced cases that are often but the outcomes of deferred treatment.

There is no doubt that a residence at the sea-side is of infinite value in a large number of cases of scrofula. As it may be supposed, the greatest advantage is observed in instances of acquired struma, in cases where the disease has developed in the purlieus of a great town, and in those patients, in fact, to whom sea breezes and our-door exercise offer the most striking possible contrast to their previous surroundings. The records of the Margate Infirmary for scrofula support this fact by a large percentage of cases of cure, and by a still larger number where a very considerable improvement has accrued. Dr. Deligny, in the thesis already referred to, enters very fully into the question of the value of sea air in scrofula; and, founding his conclusions upon the results obtained at l'hôpital de Berck, speaks enthusiastically of its good effects.*

It must be remembered, however, that sea air is not the only curative element in the sea-side treatment. There is for those patients who are not confined to bed an absolute change in their mode of living, that has no slight effect in any good result that follows.

There are a few patients (and they are but few) whose condition is often rendered worse rather than improved

---

*See also " Rapport sur les resultats obtenus dans le traitement des enfants scrofuleux à l'hôpital de Berck-sur-mer. Paris, 1866.

by a residence at the sea-side. These are for the most part certain cases with a marked phthisical tendency, many cases of eczema, some cases of strumous ophthalmia, and here and there a case of lupus.

Sea-water baths in various forms and bathing in the open sea are valuable therapeutic agents that, however, require some discretion in their use. It is unnecessary to detail here the various instructions that have been laid down from time to time for the proper administrations of these baths. Such instructions are for the most part merely the expressions of common sense and the outcomes of a rudimentary knowledge of medicine. Dr. Deligny* gives an excellent account of the chief points in connection with this subject, and the mode of conducting the bathing establishment at l'hôpital de Berck may be well taken as a model.

With regard to the general medicinal treatment of scrofula, I think it may be said that there are few diseases for which a larger number of remedies and specifics have been advised and used than have been used in scrofula. For a long while the treatment by alkalies was regarded as sovereign. They were at first considered to act by neutralizing the acrid matter that was the foundation of all strumous disease, and on the explosion of that theory it was asserted that they were very potent in dissolving and eliminating tubercular deposits. Mercury, too, was for some time a specific on account of its supposed solvent action. The salts of barium were for many years held in great esteem, probably from the fact that they were found to produce no striking evil effects, and so compared favorably with the results that often followed when mercury was freely administered. A French therapeutist† discovered that arseniate of soda was a cure for scrofula, no matter what the local manifestation (excepting bone disease). Dr. Harkin, of Belfast, asserted that in chlorate of potash was to be found a drug of remarkable efficacy in strumous disease, and from his account we gather that

---

* "Loc. cit.," p. 91.
† M. Bouchet. "Bull. de Therepeut.," vol. lix. p. 433.

"fifteen to twenty days generally suffice to heal the most extensive ulcerations of the cervical and submaxillary lymphatic glands,"* when this salt is administered.

It is to be regretted that these and many other highly advised remedies have not proved to possess the value ascribed to them by those who recommended their use.

At the present time the chief drugs used in the treatment of scrofula are cod-liver oil, iron, iodine, and certain simple tonics. Cod-liver oil should certainly occupy the first place in this list. Its use is often attended with remarkable benefit, and it seldom fails to effect some improvement at least in the condition of the patient. Before this drug is administered, and, indeed, before any prolonged treatment is commenced, it is essential that the digestive functions should be in good order. If any of the digestive troubles exist that are so common in the strumous, they may be best managed by an occasional aperient of calomel, and the use of a mixture composed of soda, rhubarb, with calumba, gentian, or cinchona bark. At the same time the patient's appetite should be regulated to meet the needs of the case. The oil should be given for a long period and in full doses. It must, of course, be immediately omitted if vomiting or diarrhœa is induced, and may be left off or taken in smaller doses during very hot weather. The best time for its administration is about half an hour after meals, as it is less likely to occasion nausea then than if taken on an empty stomach. In those cases where the drug is not tolerated butter may be given as Trousseau advises, or the mixture of butter and iodides that he has recommended.†

Niemeyer states, with a good deal of truth, that cod-liver oil acts most beneficially in the sanguine or erethic form of scrofula, but it is certainly not contra-indicated "in the fleshy, bloated patients of the torpid class," as Birch-Hirschfield maintains. The observation of the

---

* *Dublin Quarterly Journal of Medical Science.* November, 1861.
† "Clinical Lectures." Sydenham Society, vol. v. p. 91.

latter author that "on glandular tumors it seems to produce no effect whatsoever,"* is, I am convinced, ill-founded, I have seen in many cases the most striking improvement take place in gland enlargements during the use of this oil, and Grancher† speaks of the entire disappearance of such masses when the oil alone has been given.

With regard to iodine, its use has been extolled by many, especially by Lugol; but I imagine that much less faith is placed in its efficacy now than was the case when it was first extensively used. It is a drug often badly borne by patients, and can seldom be taken for any length of time. It appears to be most applicable to chronic cases that show an absence of any inflammatory reaction, and often does good in very large glandular swellings of old standing. The dose of pure iodine usually recommended for a child is from $\frac{1}{20}$ to $\frac{1}{10}$ of a grain three times a day, and it is best given combined with one grain of iodide of potassium. I think, however, that the most valuable preparation of iodine is the syrup of the iodide of iron, and I think I have been correct in ascribing very good effects to this drug in many instances, especially in early glandular disease.

Some preparation of iron is generally called for in cases of scrofula, and particularly, of course, when any anæmia exists. The most useful drugs would appear to be the saccharated carbonate, the lactate of iron, dialysed iron, and the compound syrup of the phosphate of iron. Any of these forms are well taken by children, and the last-named preparation can be very conveniently combined with cod-liver oil. Some simple tonic is of service in many strumous cases, and quinine or some preparation of bark is certainly the best to make use of, especially in cases where suppuration is active.

It is needless to observe that a vast number of mineral waters, from sea-water downwards, have been advised as serviceable in scrofula. One of the best

---

* " Loc. cit.," p. 320.
† " Loc cit., Dict. Encycl.," p. 341.

would appear to be the Adelheid spring of Heilbronn, and much has been said in praise of the waters of Kreuznach, Nanheim, La Bourboule, etc.

Local measures.—These may be considered under two heads, viz., medicinal treatment and operative measures.

1. *Medicinal treatment.*—In treating any cases of gland disease, the first indication is to remove all sources of peripheral irritation. It is useless to attempt to cure a glandular enlargement while some lesion of the surface still exists, that has perhaps not only induced the tumor, but is also maintaining it and encouraging its increase.

It is of no avail to apply pigments and ointments to enlarged cervical glands, while an opthalmia is in active progress that has caused, or is at least keeping up the lymphatic mischief. It is useless to attempt to diminish a glandular tumor in the axilla that has been subsequent to ulcerating chilblains, if those skin affections are allowed to progress unheeded. In many cases the surface lesion has been already healed, and there is no peripheral disturbance that can account for the affection of the lymphatics. Some circumstances, however, must not be too readily assumed to exist, and in every instance a most careful examination of all parts likely to be concerned must be carried out. Particularly should the mouth and pharnyx be examined. Any ulcers may be treated by chlorate of potash or by some caustic or astringent application, carious teeth should be removed, especial attention should be directed to the tonsils; and in any case where those bodies are distinctly enlarged they should be at once excised. The condition of the conjunctiva and of the nasal and auditory mucous membrane should be considered, all eruptions of the skin should be actively treated, and a careful examination made of the hairy scalp for any sources of irritation, It must also be remembered that cervical and axillary gland disease may depend—to some extent at least—upon thoracic mischief; and in cases where the clavicular region of the neck is involved this should be borne in mind.

With regard to general local measures the part should be kept as free from irritation as possible, and at a fairly equable temperature. This more particularly applies to the neck, and when gland disease exists in that situation the part should be kept covered up. All handling of the glands should be avoided as much as possible.

I might here speak of the general indications with regard to suppuration in these gland tumors, reserving, however, the detailed consideration of abscess for a subsequent paragraph. I think it may be laid down as a rule that suppuration should not be encouraged until the pus has a free exit. To allow a large purulent collection to form is, to my mind, an evidence of bad practice. In speaking of gland abscesses, I shall venture to strongly maintain the value of the earliest possible evacuation of pus. No measures should be adopted in any case that would tend to increase suppuration before there is an exit for it, but when there is a free exit for the pus then by all means let suppuration be encouraged as much as possible. I mention these matters as an introduction to the question of poulticing. Poultices are often applied to gland enlargements in a very casual and indiscriminate manner. I would say that, with scarcely any exception, poultices should not be applied to gland tumors unless the skin has yielded or been punctured, and there is thereby a free opening for the discharge encouraged by the poultice. The poulticing of inflamed gland masses while the skin is still intact merely encourages a large collection of pus beneath the integument that allows a large abscess cavity to form, and is very apt to be attended by considerable undermining.

If pus must form let steps be taken that the collection be as small as possible by the time the matter is detected and let out. For these reasons, therefore, I think that the indiscriminate use of poultices in strumous gland affections is to be condemned. There is another general point that bears upon this matter, and it is this. Occasionally the inflammatory process in these glands, or rather in the tissues about them, is somewhat active, the parts become tender, the skin hot and perhaps a little red, and yet there is no certain indication of the presence of any pus. In such cases cold evaporating lotions are to be advised. Under their use the inflammation commonly subsides, and as the swelling of the parts becomes less marked, fluctuation at one spot can perhaps be detected. In some cases of this character that I have seen suppuration would appear to have been entirely warded off by active treatment; and with the subsidence of the acuter symptoms, the gland tumor was found to have actually diminished in size.

With regard to local applications, the first drug to be considered is iodine. Not long ago iodine in the form eiher of pigment or ointment was used very extensively as an application for gland tumors.

Glandular enlargements of all kinds and in all shapes were indiscriminately painted over with iodine, and that having been done the local treatment was considered to be at an end. It would appear now that a great change of opinion has taken place upon this question, and many surgeons now discard the iodine paint and iodine ointment altogether, and condemn them as useless.

Perhaps the truth lies mid-way between these modes of practice. The local effects of iodine would appear to be those of a stimulant or irritant, and any good that it accomplishes is due, I imagine, to the increased blood supply that it encourages in the part. Thus it will be found that applications of iodine paint do positive harm in cases of commencing gland disease, and in all instances where inflammatory processes are active. That is to say, the drug adds to the mischief and intensifies it.

But although it may be injurious in these instances it is not injurious in all. There are some few gland enlargements that are benefited by iodine applications. These are very chronic gland tumors that have assumed a most indolent course, or have come absolutely to a standstill. The effect of iodine upon these masses varies according to the condition of the tumor, but in all cases the effect depends upon an improved blood supply. If the morbid change is not very far advanced this stimulus may serve to promote resolution, especially in children, but if suppuration has commenced and is remaining for a time in abeyance, the iodine may act in encouraging the process. This latter object is seldom to be desired except in certain gland enlargements in adults, when any hope of cure other than by suppuration may reasonably be abandoned. On the whole, the use of the stronger preparations of iodine is seldom called for, and is mostly limited to gland disease in adults.

The most efficient application for these affections, so far as I am aware, is the unguentum plumbi iodidi. How this preparation acts I cannot say, but I feel as confident of its good effects in many cases as one can be in any instance where more than one mode of treatment is being simultaneously carried out. The ointment should be gently rubbed into the part for some five minutes night and morning. It is not to be advised in any cases that show distinct evidences of somewhat active inflammation, nor in cases where suppuration threatens, nor, on the other hand, in cases of very recent date. Allowing these exceptions, it will be found that the application is useful in a large number of cases, and especially if the patients are children. In adults I doubt if it is of much effect. In the case of any supposed improvement from the use of this or other preparations, it must be borne in mind that many gland cases in children show a great disposition to spontaneous cure, whereas in adults the prognosis is not so favorable. Lugol and Bazin strongly recommend an ointment composed of iodine, iodide of potassium, and lard, but I have no experience of its use.

Operative measures.—Presuming that general and local remedies have been made use of without avail, there are some cases that will be specified below where an operation of some kind may be entertained. The operative measures that are the most to be recommended are excision, scooping, and cautery-puncture.

Excision.—This plan of treatment, which consists simply in making an incision over the gland mass and enucleating it from its bed, is of somewhat limited application. In any case the operation should be regarded as the last resource, especially in children in whom, as has been already stated, resolution is not uncommon. In the case of adults the operation may be more readily undertaken, provided that the local conditions be such as permit the operation. I think that the operation is applicable to three kinds of case. 1. There is only one, or at the most only two or three gland tumors. These are perfectly indolent, and have resisted all general treatment. There is an absence of any signs of active inflammation. The gland tumor is of fair size, is superficial, is throughout of equal density to the touch, and is freely moveable. I have met with several instances of gland disease of this character in children, and have removed the tumor (or in some cases more than one) in the out-patient room, and allowed the child to go home. The results have so far been most satisfactory. These masses after removal are generally found to be more or less caseous, but to present no purulent collections. 2. There are a large number of glands that have increased without any symptoms, and that have always been free from pain and without tenderness. Some might be large, and lobulated masses might be constituted, but in any instance the glands are freely moveable, and in the case of the lobulated tumors the individual glands forming those masses can be clearly made out. There are, indeed, no adhesions of any kind. These masses should be well limited, and clear of any more general but less defined gland disease in the vicinity. Such tumors are most common at the base of the neck and in the axilla, and if left alone are apt to assume very grave dimensions before they suppurate. They shell out with remarkable ease when operated upon, and I have on more than one occasion seen a porringer full of them turned out of the axilla. Microscopic examination shows them to be as a rule glands that I have described in the second division in the chapter on the histology of the disease. Indeed, one was taken from a case where an enormous number of glands were removed from the axilla by Mr. Couper at the London Hospital. The lad, the subject of the operation, did very well.

3. A single large and fairly moveable tumor is excercising injurious pressure upon some neighboring part. Or such pressure is caused by a fairly moveable mass composed of several glands that all together do not form a tumor too large to be readily removed. Such cases are very uncommon, although instances are recorded of injurious compression effected by comparatively small and isolated gland tumors. The bulk of cases where evil pressure effects are apparent are usually cases of very large and deeply adherent glandular masses that are totally unsuitable for the operation of excision.

Before disposing of this subject two questions remain to be considered. A. What is the rationale of the operation? and B. Is the operation itself a simple one?

A. In the first two instances of gland disease just given, excision is to be advised on these grounds. The affection has proved intractable to ordinary treatment patiently tried. As has been shown, these strumous glands are apt to spread locally and to infect neighboring glands, and by their timely removal such a mode of extension is prevented. In the class of case No. 2, this local extension is very evident, and is apt to lead to a grave form of scrofulous tumor. If the operation is restricted to the species of cases mentioned, there is every reason to suppose that the local disease can be entirely eradicated. If such eradication is effected, a malady is cut short that if left might lead to prolonged evil effects, to tedious suppuration with all its probable ill consequences. The gland tumors that are referred to in instance No. 2 will in time, if left alone, become matted together, and probably lead to the most intractable form of suppurating strumous glands. The clean and simple scar left after the operation compares favorably with the cicatrix commonly formed when extensive suppuration has existed. Many have urged the operation of excision on the grounds that by the removal of caseous masses the patient, is rendered less liable to general tuberculosis. I think, however, that this argument may be entirely discarded. Even if it be allowed that general tuberculisation depends upon some cheesy focus it must be admitted that out of the enormous number of patients who present caseous deposits in their bodies, the percentage of those who fall victims to diffused tubercular disease is so very small that the probability of that disease may be put out of the question. I think also, that the argument advanced by Ruehle* in favor of removing glands on the plea that such removal may prevent phthisis, is also unworthy of consideration in discussing this mode of treatment. One word as to the other aspect of this subject. Excision of glands has been objected to on the ground that the operation if successful is apt to be followed by a fresh outbreak of scrofula in some other part. What has been already said about the antagonism between strumous diseases would appear to support this proposition. But in practice no such evil consequences are found to follow. Velpeau has very rarely seen any lung troubles, for example, appear after these operations, and Gosselin maintains that when such ill results do follow they are due to an enfeebling of the patient consequent upon an extensive suppuration after the operation,† I have been enabled to watch for long periods six patients who have undergone this mode of treatment, and in none of these cases did any fresh scrofulous disease appear. A sequence of scrofulous affections, is commonly met with in severe forms of the malady, and if exception be made of some of the cases that come under the second heading it will be observed that most of the instances of gland disease suitable for excision are not among the graver examples of scrofula. The reasons for operating in the third example above given required no detailed mention.

In no instance where a proper selection of case was made have I seen fresh glandular troubles arise that could be ascribed to the operation.

B. I think this operation may be regarded as a simple one if it is restricted to the cases mentioned. There

---

* Loc. cit., Ziemssen's *Cyclop.*, vol. v. p. 605.

† Quoted by Humbert. *Des Neoplasmes des ganglions lymphatiques.* Paris, 1878, p. 138.

are, however, certain points of difficulty. In the first place, the surgeon may be very much deceived as to the mobility of the mass he proposes to excise. A gland that feels fairly moveable before the skin is incised may be found very adherent to the deeper parts when its removal is attempted. For the operation to be a satisfactory one the tumor should be loose enough to shell out easily with the handle of a scalpel after a few touches with the blade. Anything like prolonged dissection is not desirable, and a violent tearing out of glands is perhaps as bad. Indeed if some of these glands are roughly handled the capsule is apt to give way and the caseous contents to escape, thereby still further complicating the operation. In the second place, the surgeon may be deceived as to the number of diseased glands he is about to remove. Perhaps only two glands are conspicuous before the operation, but on excision these others are discovered; they are removed, and more glands deeper down still are encountered, and so on. In such cases it is better to do too little than too much. In the case of numerous gland tumors mentioned under the second heading as given above, discretion must be exercised as to how far the operation is to extend whenever it is found that the tumors reach down among the deeper parts. In these cases however, the ease with which the masses generally shell out will allow of them being safely removed from considerable depths. In one case I removed a number of glands from behind the carotid vessels, the wound was dressed antiseptically and did well. In removing deeply placed glands from the base of the neck there is much risk of injuring the dome of the pleura, to which structure I have seen these glands in more than one instance adherent. In operations of any magnitude about the neck there is of course the usual risk of hæmorrhage and the possibility of an entrance or air into injured veins. Mr. Holmes* relates the case of a child whose axillary artery he accidentally wounded

---

* *Surgical Treatment of the Diseases of Infancy and Childhood,* London, 1868 p. 639.

in removing some gland-masses from the armpit, the tumors being adherent to that vessel. When such adhesions exist it can be understood that this accident is by no means unlikely to occur, even under the hands of so experienced a surgeon as Mr. Holmes. The vessel in this instance was ligatured above and below the wound, and the case did well. In all cases where the wound is large or deep, it is well that it be dressed throughout antiseptically according to Lister's method. As a rule, healing takes place readily and well if the edges of the incision have been carefully approximated.

Scooping.—This plan of treatment is thus carried out. An incision about a quarter of an inch in length is made in the skin over the gland to be operated upon, and is then continued through the capsule of the gland itself. A Volkmann's "spoon" is then inserted, and the contents of the gland scooped out. In cases where sinuses exist the scoop may be passed through one of them, provided that the condition of the gland does not forbid the operation.

This procedure can be adopted in a good number of cases of strumous gland disease. It is *not* applicable to the various gland affections mentioned as suitable for excision. The "spoon" should not be used for any glands that are freely moveable, or that are of recent date, or that show evidence of very active mischief. The cases to which it may be most advantageously applied are these: glands that have resisted treatment and are of long standing, that have attained good size, and are either becoming soft or are distinctly softened. Especially is it important that these tumors should be adherent, and none are better suited for the operation than are those that are closely adherent to the skin. Such glands are caseous either wholly or in part, and present larger or smaller purulent foci. Some of these glandular tumors will be of great size, and made up of lobulated masses matted together. In some parts of such a mass suppuration might have occurred and sinuses be present; it may then be possible to attack the glands whose capsules are still intact through the medium of these sinuses. The operation should not be performed upon actively inflamed glands, nor upon glands that show gross evidences of suppuration. If the tumor be large it may be advisable to insert the instrument through more than one opening in the skin. The rationale of the operation is very simple. The glands operated upon are of a kind that cannot be expected to end in resolution. Natural processes would in time throw out the diseased material, but such elimination would be tedious, and attended with much local disturbance, even if it made no impression upon the general health. There would be prolonged suppuration and possibly unsightly scarring. On the other hand, the operation effects in a few minutes what natural processes would probably require months to bring about. With regard to the operation itself, it may be considered as a simple one if the cases be properly selected.

If the "spoon" be applied to glands that are freely moveable under the skin, the loose connective tissue about the tumors is opened up, and into that tissue some of the morbid products of the gland may readily escape. Thus if non-adherent glands are operated upon abscess is likely to form, and undermining of the skin and other evils are apt to follow. It happens that the glands most suitable for the treatment are those that have usually the thickest capsules, and thus there is little risk of the instrument straying beyond the tumor operated on. The capsule is of course left behind, but it gives no trouble, and probably shrinks into a harmless mass of fibrous tissue. It must be remembered that its vascular supply is often considerable when compared with that of the gland tissue itself. After the operation, it is well to gently syringe out the cavity with a weak carbolic solution, and in all cases the operation should be performed with antiseptic precautions. If a large cavity is left, a drainage tube should be inserted and maintained in position for so long as required. I have seen very good results follow in instances where no antiseptic precautions have been taken, but as in all cases there is some chance of opening up the cellular tissue of the neck, those precautions had better not be neglected.

Cautery puncture is one of the very best operative measures at the disposal of the surgeon for the cure of scrofulous glands. I had practised it myself in many cases before I was aware that it had been advised and used with success by certain surgeons in France. In this operation I make use of a thermo-cautery point about as thick round as a No. 7 catheter. This point, having been heated to a bright-red heat, is thrust through the skin into the substance of the gland, and passed in three or four directions in the body of the tumor before it is removed. If the gland be at all moveable it is necessary that it should be firmly fixed with the thumb and forefinger while the cautery is being applied. If no pus or cheesy matter follows the removal of the iron a simple zinc dressing may be applied, but if any such matters escape then a poultice should be ordered. This simple operation is applicable to a large number of glandular swellings. It may be used in any of the tumors described as suitable for the treatment by scooping, and in any of the tumors placed in the first division of the cases considered proper for excision. It will thus be seen that it is applicable to a larger variety of gland masses than is either of the two modes of operation already described. I consider it superior to the treatment by scooping, and it is certainly much more simple and much more easy to perform than is that operation. It is more adapted for adherent than for moveable glands, and should not be practiced upon tumors that are less in size than a large cherry. If the gland mass contains no pus and no caseous matter soft enough to escape, little or no discharge follows the puncture; but the tumor, after a temporary enlargement, begins to shrink, and soon terminates in cure. If any pus be present it has a free exit, and the fact of the aperture having been made by a cautery, insures its remaining patent for a considerable period.

How this procedure acts when it cures glands that contain neither pus nor softened cheesy matter, I cannot say. It obviously excites a healthier action, and leads to very satisfactory resolution.

In the case of moveable glands, the charring of the

parts that occurs along the line traversed by the cautery would appear to prevent that extension of mischief into the adjacent cellular tissue that is apt to occur when a Volkmann's "spoon" is made use of.

The time required to effect a cure under this mode of treatment varies according to the size and condition of the gland tumor. In cases where no pus or caseous matter escapes, some fourteen or twenty-one days are sufficient to bring about the cure of the mass, the puncture having healed a long while before that time. If there be much pus or broken down material in the gland a longer time may be necessary, although many cases end favorably within the period. In one or two instances a very rapid resolution took place. I have tried this simple operation some twenty times, in all cases in children, and the results have been extremely satisfactory. In only one instance did the skin become undermined, and then the mischief was but of limited extent. The scar left is simple, small, and in way conspicuous.*

The operative measures now to be briefly alluded to are, I think, of but little practical value, while several of them have already been abandoned as useless.

Interstitial Injections.—This plan of treatment was some time ago extensively carried out on the Continent, but I believe it is made much less use of at the present time. A Pravaz's syringe is used, and the point thrust well into the interior of the gland, the material to be injected being then discharged. A vast number of different solutions have been used, and among them may be mentioned, tincture of iodine, alcohol, chloride of zinc, pepsine with or without dilute hydrochloric acid, various dilute acids, and nitrate of silver. Dr. Morell Mackenzie† appears to have made the most extensive use of this treatment, and his conclusions may be summarised as follows: the treatment may be either by promoting resolution of the gland, or its destruction by

---

* A brief notice of this treatment, with cases under the care of Dr. Perier at l'hopital Saint-Antoine, will be found in the *Journ. de Med. et Chirurg.* Paris, Jan. 1881, p. 17.

† *Med. Times and Gazette*, vol. i. 1875, p. 577.

Cautery measures in scrofulous cases. To effect the former end, the dilute acetic P. is used. Five to twenty drops are ... ng to the size of the gland. The injec- ... be repeated not oftener than once a week, ... erage duration of the treatment is three ... o effect destruction of the tumor three to ... a solution of nitrate of silver ($\mathfrak{z}$ j to $\mathfrak{z}$ j) are ... Three to four injections are usually sufficient ... ly carry out this treatment. Out of twenty-three cases treated with acetic acid fifteen were quite ... nd a like good result followed in three out of ... ve cases treated with the silver solution. Dr. ... kenzie gives no details as to the proper cases to ...ct for these very different modes of treatment.

French surgeons for the most part recommend iodine for injections, using from five to ten drops of the tincture for each application. The treatment has to be repeated some five or six times in each case.*

Seton.—This treatment is applicable to large indurated gland tumors, and its mode of cure is merely by effecting supuration in the body, and thereby bringing about its elimination. It must be owned, however, that there are better curative measures at the surgeon's disposal.

A great point urged by those who favor this operation is that it leaves very little scar. A seton composed of a single thread of silk should be passed through the tumor from end to end in its long axis. In a few days the gland swells, inflames, and becomes painful. About the twelfth or fifteenth day it softens, and by the twentieth or twenty-fifth day suppuration is well established (Deligny).

Electricity has been used in various ways, but the accounts of its value are somewhat conflicting.

The remaining modes of treatment may, I think, be discarded as useless, if not as actually detrimental. They comprise crushing the gland by violent compression between the thumb and fingers through the uninjured

---

*See "Des Neoplasmes des ganglions lymphatiques," by Dr. Humbert. Paris, 1878, p. 137. See also observations and case by Dr. Marston. "Bull gen. de Therap.," 1876.

skin. The capsule ruptures and the mass, if possible, may be broken up. As may be supposed, suppuration commonly follows. Then the treatment by subcutaneous laceration of the gland by means of a cataract needle inserted beneath the skin, and lastly, the plan of treating these glands by continued compression.

A few of the commoner complications of lymphatic gland disease may now be considered.

Gland Abscess.—There are two points in the treatment of these abscesses upon which a large number of surgeons are agreed. In the first place, the purulent collection should be opened as soon as possible, as soon, indeed, as there is any evidence of pus; and secondly, the opening made for the evacuation of the matter should be as small as possible. In 1871 the editor of the *British Medical Journal* obtained from a number of hospital surgeons expressions as to their opinions about the treatment of these abscesses. The greater number were in favor of early incision and small puncture.* I can conceive of no valid arguments in favor of the practice of allowing these abscesses to break spontaneously. By such practice a large abscess cavity is allowed to form, the skin becomes extensively undermined, troublesome sinuses and ulcers usually follow, and end in unsightly cicatrices. Free incisions into glandular abscesses also are very commonly followed by like ill results, and are certainly to be condemned. With regard, however, to abscesses that have formed in the connective tissue outside a gland, the capsule of which is still intact, some reservation must be made. These collections are not difficult to diagnose from those within the glands by means of the signs already given. At the bottom of such collections a diseased gland is commonly to be seen, and until this body has been removed, either by natural or artificial means, the suppuration is likely to continue. It is well, however, to make a rule of opening these abscesses by a small puncture, for under such treatment the case may do well. If the suppura-

---

* *British Medical Journal*, vol. ii. 1871, p. 727, et seq.

tion continues the incision can at any time be enlarged, and the exposed gland treated. It is rarely advisable to attempt either to enucleate this body from its bed or to dissect it out. It is usually very adherent, and the adjacent parts much inflamed. The best plan is to thrust a point of the thermo-cautery into the gland, and allow it to shrink or to discharge its contents. I think this treatment is better than that advised of destroying the gland with potassa fusa, or of dusting it over with the red oxide of mercury.

As to the best mode of evacuating gland abscesses it will be observed that most of the plans of treatment recently advocated have these common features—a small opening, and no handling of the gland, after the pus has been let out. Thus Sir James Paget* advised a small puncture some two lines in length, with care to avoid any pressure upon the part. Mr. Lawson Tait† went a step further, and recommended that the pus should be drawn off by repeated punctures with a hypodermic syringe. The same principle underlies Guersant's plan of evacuating these gland abscesses by a single seton thread, and the mode of treatment also advised by some of puncturing the body with a trocar.

I would, however, most strongly advise that these abscesses should be opened by a single puncture of a fine thermo-cautery point.

I made a careful trial of some of the principal methods of treatment advised, and found none to equal the use of the cautery. The operation takes but a moment, and I think it is but little more painful than incision with a knife. As the pus escapes no pressure should be exercised upon the part, but the matter allowed to spontaneously trickle out. I think the next best plan of opening these abscesses is by a small puncture with a tenotome, and subsequent drainage with a small india-rubber tube.

Dr. Sydney Ringer‡ has strongly advocated the use of the sulphides of potassium, sodium, or calcium in

---

* *Med. Times and Gazette*, vol. i. 1856, p. 5.
† *British Medical Journal*, vol. i. 1871, p. 117.
‡ *Lancet*, vol. i. 1874, p. 264.

SCROFULOUS LYMPHATIC GLANDS. 179

cases of glandular abscess. He asserts that these drugs often appear to arrest suppuration, or when pus has formed they hasten the maturation of the abscess, render it more circumscribed, and promote a healthy condition of the discharges. For children he advises from $\frac{1}{10}$ to $\frac{1}{2}$ grain or a grain of the sulphide of calcium every two or three hours.

Sinuses are apt to follow npon abscess in connection with strumous glands, and are often very intractable. The plans of treatment available vary greatly according to the nature of each case. Often there is not free vent for the discharge, and then the aperture may be enlarged or the abscess cavity more carefully drained. More commonly the persistence of the sinus depends upon the unhealthy action going on in the part. In such cases injections of carbolic acid lotion or of weak solutions of iodine or of nitrate of silver may be tried, and this may be combined with gentle pressure by a well-adjusted pad in cases where such compression is possible. If several sinuses near together are connected with one another by undermined integument, the skin so affected may be slit up and the cavity dressed from the bottom. Iodoform ointment is an excellent application in these and like cases.

Often the defective healing depends upon undermined skin. This skin soon becomes sodden or thinned, and is always purplish and unhealthy looking. Under such circumstances it should be without doubt destroyed, as has been repeatedly advised. This destruction is best effected by the actual cautery. It is much more ready, more certain, and on the whole less painful than the treatment by potassa fusa and other caustics. Excision of the undermined skin is never desirable. In all these cases of intractable suppuration and persistent sinuses one measure should always be observed, and that is to kept the part at rest. This applies especially to mischief in the cervical region. Few parts of the body are more constantly in movement than is the neck; and yet without taking any precautions to insure that rest that is so indispensable, the surgoon is surprised that a suppurating district in a child's neck declines to heal in spite of all his treatment. In any case where the healing process is disposed to be tardy, I apply a stock of gutta-percha to the neck, and insure an absolute quiet for the inflamed and irritated parts. This stock is readily made, and should have its fixed points above at the lower jaw and occiput, and below at the chest and shoulders. If it be well moulded over each shoulder it merely requires a circular band of strapping to maintain it in position.* Padding should be applied at all points where the edge of the stock comes in contact with the skin, and if this be done the child can wear the apparatus night and day for weeks without much inconvenience. It can be readily removed when needed to dress or inspect the wound. I have been surprised at the immense improvement that has followed upon the use of this simple collar of gutta-percha in cases that, before its use, had appeared most intractable, and feel sure that its more extended application would do away with a good many of the more obstinate cases. Sometimes they may be sore at points where the stock should press; in such cases the head and neck should be fixed by the apparatus advised by Sayre for cervical spine disease, the trouble of applying such an apparatus being well repaid by the good result that follows.

Scars.—The treatment of the unsightly cicatrices that often follow after scrofulous gland disease demands but a brief notice in this place. Any nipple-like projections that exist about the scar may be snipped off. Prominent bars and ridges often disfigure the cicatrix, and these may sometimes be removed with scissors, and some improvement thereby effected in the appearance of the part. Now and then a scar that is non-adherent, but that is unsightly from its persistently purplish color or its irregularity of surface, may be convieniently excised, provided that the adjacent parts be

---

* I first take a rough model of the neck with a sheet of paper, and then cut the stock out of a piece of gutta-percha, such as is used for making splints. This is placed in warm water until it is soft enough to be moulded to the neck. A slit here and there along the edge of the stock is needed to ensure a good fit. The vertical free borders of the stock should overlap in front in the middle line of the neck.

perfectly sound and the patient in good health. By this means, if good healing occurs, a clean white linear scar takes the place of the unpleasant scrofulous cicatrix.

The presistence of many of the scars in struma and a good deal of their unsightliness depend upon their being adherent to subjacent parts. To remedy the deformity produced by this class of cicatrix, Mr. William Adams has proposed a very ingenious operation.* The steps of the operation are as follows:—1st. All the deep adhesions of the scar are divided subcutaneously by a tenotome introduced beyond the margin of the depressed tissue. 2nd. By a little manipulation the cicatrix is everted, or, as it were, turned inside out, so that the scar tissue is made very prominent. 3rd. Two hare-lip pins are passed at right angles to one another through the base of the cicatrix, so as to maintain it in its raised and everted position. 4th. On the third day the needles are removed, and the scar tissue, now swollen and succulent, is allowed to return to the proper level of the skin. Speaking from a knowledge of cases that were operated upon several years before the publication of his paper, Mr. Adams asserts that the depression of the scar does not recur, and that the appearance of the part is considerably improved.

---

* *British Medical Journal* vol. i. 1876, p. 534.

www.ingramcontent.com/pod-product-compliance
Lightning Source LLC
Chambersburg PA
CBHW032155160426
43197CB00008B/921